Power in Megaproject Decision-making

This research-based book critically examines the complex interplay of power relations in decision-making in megaprojects. This book is tailored for academics, students, practitioners, clients, and policymakers and addresses a significant gap in understanding how formal and informal power influences decision-making.

Through applying governmentality theory and narrative inquiry from thirty-nine interviews, this book presents over sixty-five detailed stories offering a series of real-life narratives that expose the diverse forms of power at play in megaprojects. The stories reveal how deviations from norms can significantly impact project outcomes. This book uncovers the complex web of power relations, institutional norms, and individual actions driving decisions.

This book provides a framework illustrating four behavioural contexts, entrepreneurial, box-ticking, deception, and recklessness, as manifestations of megaproject decision-making, providing readers with a tool to understand and navigate the intricate power dynamics in megaprojects.

Readers will benefit from practical insights and strategies for fostering conducive decision-making environments, tailoring interventions to promote constructive behaviours, and mitigating harmful actions. This book is a resource for anyone seeking to grasp the complexities of megaproject management in the context of power relations, shaping their decision-making leading to success.

Dr Jessica Pooi Sun Siva is an award-winning researcher and academic. Jessica is currently a Lecturer in the School of Architecture and Built Environment, University of Newcastle. Jessica's research expertise lies in examining built environment relationships, networks, and systems with a particular focus on decision-making behaviour and practices. Jessica actively researches and teaches in areas of megaproject decision-making, design and project management, women participation and leadership in construction, sustainable built environment practices, circular economy, innovative pedagogies, transformational leadership, and infrastructure development. Jessica regularly engages in international research and teaching collaborations. Jessica is a Fellow of the Advance Higher Education Academy. Jessica is currently the Program Convenor for the Bachelor of Construction Management (BCM) program.

Associate Professor Thayaparan Gajendran's research focuses on the governance of organisations using sociological lenses in construction and disaster management. His research is committed to developing evidence-based practices and policies for building resilience at the complex intersections between communities and organisations. He engages in extensive international research collaborations and has published over 150 academic articles. He has received government and industry research funding of more than $1.5M. A/Professor Gajendran received two Vice Chancellor's Teaching Citations, an Australian Learning Teaching Council Citation award, two Pro-Vice-Chancellors Higher Degree Supervision awards and a Vice Chancellor Inspiring Excellence Award. A/Professor Gajendran has held numerous service positions, including Assistant Dean of Education, Program Convener of Construction Management and Disaster Management programs, and Faculty of Engineering and Built Environment Ethics Committee Chair. He is currently the Acting Head of the School of Architecture and Built Environment.

Spon Research

Publishes a stream of advanced books for built environment researchers and professionals from one of the world's leading publishers. The ISSN for the Spon Research programme is ISSN 1940-7653 and the ISSN for the Spon Research E-book programme is ISSN 1940-8005.

Successful Development of Green Building Projects
Tayyab Ahmad

BIM and Construction Health and Safety
Uncovering, Adoption and Implementation
Hamed Golizadeh, Saeed Banihashemi, Carol Hon and Robin Drogemuller

Digitalisation in Construction
Recent Trends and Advances
Edited by Chansik Park, Farzad Pour Rahimian, Nashwan Dawood, Akeem Pedro, Dongmin Lee, Rahat Hussain and Mehrtash Soltani

Managing Projects with Smart Technologies
Developing Technological Competency for Project Managers
Bon-Gang Hwang, Jasmine Ngo and Hanjing Zhu

Innovations, Disruptions and Future Trends in the Global Construction Industry
Edited by Temitope Omotayo, Temitope Egbelakin, Olabode Ogunmakinde, and Adebayo Sojobi

Power in Megaproject Decision-making
A Governmentality Approach
Jessica Pooi Sun Siva and Thayaparan Gajendran

Power in Megaproject Decision-making

A Governmentality Approach

Jessica Pooi Sun Siva
and Thayaparan Gajendran

Routledge
Taylor & Francis Group
LONDON AND NEW YORK

First published 2024
by Routledge
4 Park Square, Milton Park, Abingdon, Oxon OX14 4RN

and by Routledge
605 Third Avenue, New York, NY 10158

Routledge is an imprint of the Taylor & Francis Group, an informa business

British Library Cataloguing-in-Publication Data
A catalogue record for this book is available from the British Library

ISBN: 978-1-032-45095-7 (hbk)
ISBN: 978-1-032-45545-7 (pbk)
ISBN: 978-1-003-37748-1 (ebk)

DOI: 10.1201/9781003377481

Typeset in Times New Roman
by codeMantra

This book is dedicated to:
Bernard and Oskar
Gopi, Tanishkaa, and Mohniishen

Contents

Figures

Tables

Boxes

Preface

The exploration of decision-making in megaprojects presents a compelling and complex challenge, given the transformative impact these projects have on cities and economies. Megaprojects, characterised by their large budgets, long durations, and intricate complexity, often become arenas for power dynamics that critically influence both the decision-making and overall performance of these projects. Yet, understanding the actual decision-making practices within these colossal projects remains an area with significant knowledge gaps. The frequent failures in megaprojects often lead to calls for more stringent frameworks and controls. However, this approach can sometimes be counterproductive, as it may not address the underlying issues of power dynamics and informal decision-making practices.

The scale and complexity of megaprojects almost guarantee that not everything will go according to plan. Deviations and non-conformities are common, and the capacity to adapt and respond to these unforeseen changes is crucial. Beyond the formalised contracts and established protocols lies the realm of informal governing – the practical, on-the-ground decision-making that often deviates from official plans. This includes tacit knowledge, spontaneous and experience-based decisions, and interpersonal dynamics that play a pivotal role in guiding projects through challenges. In megaprojects, power is not just about formal authority or hierarchical structures. It encompasses a wider spectrum of influences, including informal networks, personal charisma, expertise, and the ability to sway opinions and outcomes. These aspects of power often operate behind the scenes, subtly shaping decisions and project trajectories.

The problem, therefore, centres on the challenge of understanding and capturing the diverse forms of power at play in megaproject decision-making. This issue is critical because these forms of power directly influence how deviations are managed and resolved within the megaproject environment. Existing models and frameworks often fail to address these aspects, leading to a gap in our understanding of how decision-making actually unfolds in the complex megaproject environment.

Addressing this knowledge gap involved diving deep into empirical research to unravel the nature of power relations that influence decision-making in megaprojects. This research-based book moves beyond the idealised and controlled perspectives of decision-making, delving into the more nuanced, dynamic, and often informal aspects of power and influence. Adopting governmentality theory, this book provides a series of detailed and real-life narratives to reveal the different ways in which power is exercised, negotiated, and manifested in megaprojects. It presents valuable lessons on the role of deviant practices in influencing decisions, the conditions that foster deviant behaviour, and how it impacts the project's outcomes. It sheds light on the implicit rules and norms that govern behaviour, revealing how they dictate interactions, negotiations, and decisions among project stakeholders.

By grounding the study in critical realism, this book goes beyond mere description, uncovering the causal factors driving decision-making. This approach is particularly pertinent in the context of megaprojects, where decisions are often the outcome of intricate power relations rather than straightforward technical or procedural considerations. The use of narrative inquiry as a methodological approach was particularly effective for this type of investigation. By collecting and analysing detailed stories from clients and project team members, this book gains deep insights into how decisions are made, revealing the often hidden or unacknowledged factors that influence these processes. Sixty-five detailed stories of practice were gathered through 39 interviews, offering a diverse and comprehensive perspective on decision-making in megaprojects. The analysis of these stories through the lens of governmentality theory, critical realism, and narrative inquiry enabled a deeper understanding of the role of power and deviance in shaping decisions. This approach not only identified what decisions were made but also elucidated the underlying power dynamics and motivations that drove these decisions.

The focus on deviance within decision-making processes is particularly significant, highlighting how deviations from standard or expected practices can have a profound impact on project outcomes. By connecting different stories and examining the patterns and themes that emerge, the book paints a comprehensive picture of decision-making in megaprojects. It provides valuable insights into the complex interplay of power relations, institutional norms, and individual actions, offering a deeper understanding of how decisions in such large-scale and complex projects are made.

The governmentality framework for megaproject decision-making introduced in this book is instrumental in delineating the diverse forms of power, both overt and covert, that are actively exercised in the day-to-day management of megaprojects. Central to this framework is the identification of four distinct behavioural contexts that arise from the interplay between formal governance and informal governing practices, namely, entrepreneurial, box-ticking, deception, and recklessness. Case studies have been presented as examples to demonstrate how clients and project team members navigate the decision-making environment in each behavioural context. The governmentality framework is a powerful tool in preparing megaproject clients and team members to better position themselves and strategically intervene decision-making processes to steer projects towards more favourable outcomes. This book is particularly valuable for practitioners, policymakers, and academics in the field of megaproject management. The framework is designed to help predict behavioural patterns in megaprojects, thereby allowing clients and project teams to exert greater control over project outcomes. Moreover, the rich examples of the different types of deviance and their potential impacts offer practical insights and strategies for cultivating decision-making environments and tailoring interventions to encourage constructive behaviours and mitigate potentially harmful actions. The governmentality framework provides a lens through which the intricate web of power relations and their impact on decision-making can be understood and navigated, making this book a critical resource for anyone involved in or studying the field of megaproject management.

Synopsis

The quest for improved performance in megaprojects demands a paradigm shift in understanding the decision-making behaviours of clients and project team members. Historically, the focus has been on the "black box" of megaproject management, emphasising the prescription of ideal behaviours through structured governance models. To improve megaproject performance, there is a need to delve deeper into the "black box" to understand the actual practices, relationships, and power struggles that occur in the decision-making environment, beyond what is prescribed on paper. Formal governance policies and procedures provide a framework for predicting outcomes and assessing the likelihood of success, but they are not definitive. Therefore, outcomes in megaprojects depend not just on the application of rules but also on the interplay between governance (the rules and structures) and governing (the actual behaviour and actions of the participants). The governmentality approach introduced in this book marks a significant shift in understanding megaproject management by simultaneously considering formally prescribed governance mechanisms and the often-underappreciated informal power relations that influence decision-making. This dual focus is premised on the understanding that power dynamics, both overt and covert, significantly impact the execution of formal governance frameworks in megaproject environments.

To begin, this book presents a comprehensive review of the megaproject discourse, highlighting shortcomings in existing models and advocating for a focus on the study of power relations and everyday decision-making practices in these complex projects. This is succeeded by the exploration of governmentality theory to understand the complexity of decision-making in megaprojects, examining the various forms of power exercised in daily practices by clients and project teams. Additionally, insights from deviance literature are incorporated to further understand the diverse ways power is exercised in megaprojects. A governmentality framework for megaproject decision-making is presented to demonstrate how various combinations of governance and governing factors in megaprojects lead to specific management challenges, necessitating tailored mechanisms to encourage constructive deviance and mitigate destructive deviance. Multiple stories and case studies that exemplify deviations common in megaproject decision-making are presented, illuminating the often hidden and routine practices used to navigate and sometimes challenge the formalised structures and protocols of megaprojects.

The stories highlight how specific conditions could lead to both destructive and constructive forms of deviance. In summary, this book demonstrates that adopting a governmentality approach to megaproject decision-making acknowledges the importance of informal governing practices, enabling a deeper analysis of deviance and its impact on enhancing project performance. Despite the inherent uncertainties in megaproject decision-making, preparing clients and team members through understanding the use of deviance can lead to more favourable outcomes. By identifying key factors that facilitate deviance, specific governance frameworks and intervention mechanisms can be tailored to enhance predictability and control in decision-making.

Abbreviations

AEC	Architecture, Engineering, and Construction
AHP	Analytical Hierarchy Process
APM	Association of Project Management
CDC	Client Design Consultant
CEO	Chief Executive Officer
CRGP	Collaboratory for Research on Global Projects
COST	European Cooperation in Science and Technology program
DCA	Department of Civil Aviation
DO	Design Order
FMADM	Fuzzy Multi-attribute Decision-Making
FTOPSIS	Fuzzy Technique for Order Preference by Similarity to Ideal Solution
ICAO	International Civil Aviation Organisation
LTA	Land & Transport Authority
MHA	Ministry of Home Affairs
MoD	UK Ministry of Defence
MRT	Mass Rapid Transit
OGC	UK Office of Government Commerce
PPP	Public Private Partnership
QGC	Quebec Governance Framework
CDF	Civil Defence Force
SPV	Special Purpose Vehicle
URA	Urban Redevelopment Authority

1 Introduction – setting the scene

Background to megaprojects and their importance

Megaprojects, encompassing vast, multi-faceted undertakings often spearheaded by temporary collaborations among architectural, engineering, and construction firms from various countries, have become an increasingly prominent feature of the modern construction landscape (Denicol et al., 2020). These projects are not just monumental in scale but also in their implications. They are intricately woven into the fabric of society, playing significant economic, social, and symbolic roles (Flyvbjerg, 2014; Klein & Aubry, 2017).

The impact of megaprojects extends far and wide, influencing large populations through both tangible and intangible outcomes. They are pivotal in delivering essential services, advancing public health initiatives, enhancing transportation and logistical infrastructures, and promoting environmental sustainability (Klein & Aubry, 2017). The success or failure of such projects, therefore, has far-reaching consequences, affecting countless lives and shaping the trajectory of communities and nations.

However, the benefits that megaprojects are purposed to deliver are increasingly under scrutiny. Despite their recognition as potent instruments of economic growth, voices in academia and industry alike have begun to question the assumed advantages of these projects (Altshuler & Luberoff, 2003; Merrow, 2011; de Bruijn & Leijten, 2008; Jia et al., 2011, Novy & Peters, 2012).

Challenges of megaproject management

Megaprojects – which cross over both public and private sector domains – are infamously known for their long record of poor delivery (Flyvbjerg, 2005; Li & Guo, 2011). These large-scale projects are notorious for their difficulties in execution. Despite the recent surge in the number and scale of megaprojects being undertaken globally, their performance often falls short of expectations, particularly concerning economic efficiency, environmental impacts, and garnering public support (Brookes, 2012; Merrow, 2011; Flyvbjerg et al., 2003; Miller & Lessard, 2000).

The megaproject environment, while offering a plethora of opportunities for innovation, development, and growth for both clients and project teams, is also

DOI: 10.1201/9781003377481-1

fraught with considerable risks. These projects are characterised by their immense scale and scope, involving complex technological requirements and a myriad of social considerations. The sheer size and complexity of megaprojects necessitate a high level of strategic behaviour and planning (Altshuler & Luberoff, 2003; Flyvbjerg et al., 2003).

Megaprojects represent a fascinating and complex interplay of organisational dynamics and power relations (Clegg & Kreiner, 2013). These large-scale projects are deeply embedded in a social fabric shaped by the power dynamics and interactions of a diverse network of individuals and groups. Each participant brings their unique interests, values, and rationalities to the project, contributing to a pluralistic environment where power is distributed among various team members and stakeholders (van Marrewijk, 2015; Ninan et al., 2019; Biesenthal et al., 2018).

The complexity of megaprojects are sites of conflicting institutional logics. These conflicts arise from the differing priorities and perspectives of the numerous stakeholders involved in a megaproject. As a result, power in these projects is not centralised but diffused across a wide range of actors, each wielding influence over different aspects of the project. The stakes in megaprojects are extraordinarily high. The immense financial investments, job creation potential, environmental impacts, public scrutiny, and national prestige associated with these projects amplify the strength of interests and power relations involved. This high-stakes environment contributes to the unique decision-making complexities and challenges that characterise megaprojects.

Decision-making in megaprojects transcends the realm of technical processes, encompassing a broader spectrum that includes political, institutional, and psychosocial factors (Biesenthal et al., 2018; Pinto, 1996; Williams & Samset, 2010; Pitsis et al., 2018). The unique nature of megaprojects characterised by their vast scale and complexity – or "mega-ness" – intensifies the organisational and institutional challenges. This complexity gives rise to an intricate web of interdependencies among various aspects of the project, often leading to conflicting agendas and a multitude of complex problems that require multi-faceted solutions (Steen et al., 2017; Brady & Davies, 2014). These challenges are akin to what Crawford and Pollack (2004) describe as features of "soft" projects. They necessitate a sophisticated approach to decision-making that goes beyond traditional methods.

Efforts to improve megaproject performance

Since the 1950s, the quest for enhanced performance in megaprojects has been a focal point for academics, industry practitioners, and policymakers, with a particularly intensified focus in the past decade (Altshuler & Luberoff, 2003; Klein & Aubry, 2017). This pursuit has branched into three primary avenues of exploration, each addressing a fundamental cause of the challenges inherent in megaprojects: strategic rent-seeking, governance regimes, and organisational complexity (Sanderson, 2012).

1 *Strategic rent-seeking*: The first path focuses on strategic rent-seeking as a key reason for poor performance, specifically cost overruns. Researchers in this

domain argue that these issues stem from opportunistic and deceptive behaviours. They suggest that enhancing accountability through robust mechanisms is a critical solution to mitigate such practices (Flyvbjerg, 2005, 2009; Flyvbjerg et al., 2002, 2003; Bruzelius et al., 2002; Wachs, 1989).

2 *Governance regimes*: The second path investigates the governance aspect of megaprojects. Here, the contention is that many problems arise due to misaligned or underdeveloped governance regimes that struggle with the inherent turbulence of megaprojects. The advocated solution focuses on improving governability through the development and implementation of more effective governance mechanisms (Miller & Lessard, 2000; Morris, 2009).
3 *Organisational complexity*: The third avenue of research delves into organisational complexity. This perspective views the performance issues as arising from the inevitable complexities, ambiguities, and conflicts typical of megaprojects. The proposed solution here lies in developing mechanisms to manage and simplify these complexities, making them more navigable and less disruptive to project outcomes (van Marrewijk et al., 2008; Pitsis et al., 2018; Clegg et al., 2002; Clegg & Kreiner, 2013).

Despite their distinct perspectives, the three principal approaches for enhancing megaproject performance converge on a key shared assumption: the concept of actor farsightedness. This notion presupposes that effective project governance can be proactively designed with foresight, effectively planning for future scenarios before they unfold (Sanderson, 2012). Under this paradigm, the root causes of project challenges are often attributed to inadequate planning and administration. The path of positive change, therefore, is perceived to lie in the formulation and execution of comprehensive, well-conceived plans and policies, grounded in rational and democratic principles. However, this idealised model of rational, foresighted planning has been increasingly questioned by recent studies, highlighting that megaprojects often deviate from this rational model (Brunet & Aubry, 2018; Miller & Lessard, 2008; DeBlois et al., 2011; Flyvbjerg, 2009; Gil, 2009; Clegg et al., 2002).

The interplay of power and politics is a fundamental aspect of megaprojects, profoundly influencing their trajectories and outcomes (Clegg et al., 2017). These projects are arenas where power is exercised through a combination of formal structures and informal networks, significantly shaping the multitude of decisions that guide their course (Biesenthal et al., 2018; Salet et al., 2012).

Despite advancements in megaproject research across various disciplines, a critical gap persists: a comprehensive understanding of the intricate power dynamics at play in the decision-making processes of these projects. This gap is particularly evident in the lack of detailed accounts that delve into how power relations impact decision-making as projects evolve and develop (Soderlund et al., 2017; Biesenthal et al., 2018). The challenge lies in bridging the divide between the theoretical, often idealised notions of how decision-making in megaprojects should occur, and the complex, often messy reality of how it actually unfolds. This implies a shift from theoretical models towards a more empirical exploration of megaprojects, capturing the lived experiences, power struggles, and political manoeuvrings that characterise these projects.

The knowledge gap in understanding megaproject management

The existing governance frameworks and models for megaprojects often overlook the critical role of power relations in influencing decision-making. There is a growing recognition that enhancing megaproject performance necessitates a fundamental shift in the decision-making behaviours of clients and project team members. However, our understanding of the realities of such governing practices remains limited. There has been an excessive focus on prescribing ideal behaviours for managing megaprojects, akin to "looking at the black box of megaproject management" without truly understanding its internal dynamics (Soderlund et al., 2017, p. 9).

The metaphor of the "black box" in this context refers to the internal workings of megaproject management, which are often opaque or unexplored in existing research. There is limited insight into the diverse forms of power exercised in the practical implementation of megaprojects (Volden & Andersen, 2018; Brunet, 2018; Clegg & Kreiner, 2013). These forms of power, both formal and informal, play a pivotal role in shaping the outcomes of these projects but are frequently overlooked in conventional governance models. This situation suggests that the solution to improving megaproject performance might not lie in creating another idealistic or prescriptive model (Salet et al., 2012).

Decision-making in megaprojects extends beyond the confines of formalised contracts and established procedures, often being shaped significantly by spontaneous uses of power and other informal mechanisms (DeBlois et al., 2011; Ninan et al., 2019). The nature of megaprojects inherently involves deviations from planned processes (Morris & Hough, 1987; Fiori & Kovaka, 2005). Teams formed for these projects frequently focus on the practical goal of "getting the job done", which can sometimes mean diverging from rigidly defined protocols (Hallgren & Wilson, 2007). This reality highlights the importance of gaining a comprehensive understanding of megaproject management that acknowledges both the formal structures and the informal dynamics at play. Traditional frameworks and methodologies, while essential, only provide part of the picture. The less visible but equally influential informal power relations and unplanned strategies employed by project teams are critical components that drive actual decision-making in megaprojects.

The gap between the theoretical development of governance frameworks and their practical application in megaprojects is a significant issue in project management research. While there has been considerable focus on creating and refining these frameworks, the actual process of translating these frameworks into actionable practices in the real-world context of megaprojects has not been thoroughly examined (Brunet & Aubry, 2018). A key area that requires further exploration is the role of social and political skills in the governance of megaprojects. These skills, which include the ability to mobilise resources, negotiate among diverse stakeholders, and manage conflicting interests, are increasingly recognised as crucial for successful project governance, often more critical than technical evaluation skills (Morris, 2013; Winch & Leiringer, 2016).

An alternative approach to megaproject decision-making

An alternative approach exploring both formal and informal mechanisms in megaproject decision-making is crucial due to the influential role of power relations in shaping the execution of formally established governance frameworks. While formalised, legitimate forms of power provide a structural foundation for day-to-day operations, they coexist with more subtle, often covert forms of power. These less visible elements include political manoeuvres and strategic calculations that play a critical, and sometimes more significant, role in determining the trajectory and outcomes of megaprojects.

This juxtaposition of overt and covert power forms highlights the complexity of decision-making in megaprojects. The overt power is seen in the adherence to established protocols and guidelines, forming a visible layer of governance. In contrast, covert power operates through nuanced, often discreet practices such as political negotiation, strategic alliances, and behind-the-scenes manoeuvring. These practices are not always apparent but are pivotal in shaping the course of megaprojects.

This book aims to bridge the critical research gap in the field of project management by focusing on the intricate dynamics of power that play out in the decision-making processes of megaprojects. These dynamics are deeply rooted in both formal and informal social relationships, institutionalised rules, shared meanings, and cultural norms. Drawing inspiration from seminal works such as Morris and Hough's (1987) influential publication – "The Anatomy of major projects" and further developed by scholars like Cicmil et al. (2006) and Winter et al. (2006), this book aligns with the growing projects-as-practice field. This perspective shifts focus from macro-level abstractions to the micro-processes that underpin the translation of governance frameworks into practical action in the real world of megaprojects.

This book seeks to make three significant contributions to the existing body of knowledge:

1. *Exploring actual power relations and deviations*: It delves into the real-life power dynamics and deviations experienced by clients and project team members as they navigate and enact prescribed governance frameworks in the context of actual megaprojects. By examining these dynamics, the book aims to reveal how formal governance structures are interpreted, adapted, and sometimes circumvented in practice.
2. *Theorising power in decision-making*: The book endeavours to elucidate the megaproject decision-making environment through various theories of power. It seeks to understand how power is created, distributed, and exercised within the complex milieu of megaprojects, moving beyond traditional models to encompass a more realistic portrayal of power dynamics.

3 *Focusing on micro-level, "everyday" decision-making practices*: Instead of concentrating solely on high-level, macro decisions, this book pays particular attention to the "everyday" decision-making practices that often go unnoticed or are taken for granted. These micro-level practices, occurring in the day-to-day operations of megaprojects, are instrumental in shaping project outcomes and are a vital component of understanding project governance in a comprehensive manner.

The concept of governmentality

The incorporation of Michel Foucault's concept of governmentality into the study of megaproject governance offers a profound shift in understanding the intricate dynamics of power and governance in these large-scale projects. Developed in the 1970s, governmentality extends beyond traditional notions of governing by encompassing not just the processes and structures of governance, but also the underlying rationalities, tactics, strategies, and the myriad ways individuals and groups exert influence and control over others and themselves (Foucault, 1991; Boyce & Davids, 2004; Dean, 2010).

A distinctive aspect of Foucault's governmentality is its focus on the art of governing, capturing the nuances of how individuals and organisations lead, manage, and direct within the complex environment of megaprojects. This perspective explores the depth of rationalities, desires, and aspirations that drive decision-making processes, offering insights into the often subtle ways in which power is exercised.

One of the critical contributions of this perspective is its emphasis on going beyond conventional governance models, seeking to develop a more nuanced understanding of power, one that informs an emancipatory politics rooted in concepts like self-surveillance and subjectification (Foucault, 1986; Ninan et al., 2019). The strength of governmentality lies in its ability to delve into the "messy realm" of real-world practices and power relations. It sheds light on how governance mechanisms are often enacted in compromised, corrupted, and partial ways, revealing the complexities and challenges of managing and navigating megaprojects.

Stories of deviance

The exploration of deviance in the context of megaprojects offers a compelling perspective on understanding power and decision-making in these complex endeavours. Rooted in a century of management research, the study of deviant behaviour in organisational settings has delineated two primary categories: constructive and destructive deviance. Constructive deviance, as the name suggests, pertains to actions that, while deviating from the norm, contribute positively to the organisation or project. In contrast, destructive deviance refers to behaviours that have negative consequences (Warren, 2003; Bodankin & Tziner, 2009).

Recent studies have begun to challenge the traditional view that equates deviance solely with negative outcomes. For instance, research points to the potential benefits of certain forms of deviant behaviour in collaborative environments, suggesting that deviance can, in fact, be constructive and beneficial for projects (Memon et al., 2021; Chan & Jia, 2019; Bridge, In & Rowlinson, 2015). This emerging evidence opens up new avenues for understanding the dynamics of megaprojects, particularly in terms of how team members exercise power and make decisions.

This book "unlocks" the black box of megaproject decision-making, revealing the inner workings of megaproject decision-making environments. Key to this exploration is an in-depth examination of the deviant tactics employed within megaprojects. These tactics, often employed for coping with and navigating the complex, multi-faceted challenges inherent in establishing and managing such large-scale projects, are crucial to understanding how decisions are made, modified, and manipulated.

The task at hand involves two key objectives:

1 *Identifying deviant tactics in megaprojects*: This involves a careful examination of the various ways in which team members on megaprojects exercise their power in everyday decision-making practices. By identifying these tactics, it becomes possible to understand the nuanced ways power is wielded and decisions are influenced in the megaproject environment.
2 *Distinguishing between constructive and destructive deviance*: This classification is crucial in determining the nature of the deviance and its overall impact on the project. Understanding the positive aspects of certain deviant behaviours can be instrumental in harnessing these actions for the benefit of the project.

A note on method

This book presents findings from case studies of two significant megaprojects, a sports facility infrastructure and an airport in Asia, exploring the complexities of decision-making and governance in these large-scale projects. The first megaproject, noted as the world's largest sports facility infrastructure and the country's first and largest Public–Private Partnership project, and the second megaproject, recognised as the world's largest low-cost carrier airport terminal, offer rich contexts for examining megaproject management.

Thirty-nine interviews, each lasting from 30 to 180 minutes, were conducted with diverse participants from these projects. The narrative inquiry approach guided the interview process, focusing on open-ended questions to elicit detailed, personal stories from participants. The interviews explored three main areas: the participants' roles in the megaproject organisation, their understanding of the project governance mechanisms including policies and procedures, and their experiences with deviations in daily operations and decision-making, especially in instances where standard governance approaches failed to resolve issues effectively.

This book presents a collection of stories from the two megaprojects, highlighting the deviations and unexpected challenges encountered in the decision-making processes of these large-scale endeavours. These stories, which emerge from situations that could not be anticipated or planned for, focus on events that required corrective action, offering insights into the practical management of deviations in megaprojects. The compilation of these stories utilises a narrative analysis technique, where each story forms a distinct type of discourse production. Events and actions are woven into a coherent whole, structured around a plot or thematic thread. This approach enables the understanding, description, and organisation of the actions and events in the context of deviations or deviant behaviour.

During the initial phase of analysis, transcripts from the two megaprojects were examined to identify everyday events, activities, and experiences related to acts of resistance, manipulation, or deviance. This process involved using abductive logic to pinpoint critical moments or experiences that stood out as problematic in participants' daily operations and relationships within the megaprojects. Special attention was given to identifying "silent moments" where conventional protocols failed to resolve issues, leading participants to exercise deviance. The story analysis technique was employed to connect various stories from various participants, forming a comprehensive understanding of how decisions were made through the exercise of deviance.

The narratives were further analysed to distinguish between different types of deviant tactics employed by project team members. Using the paradigmatic cognition method of narrative analysis, these tactics were categorised as either constructive or destructive deviance, based on the behaviour and actions described in the stories. This classification was aligned with the primary characteristics and outcomes of constructive and destructive deviance, as detailed in Chapter 3. However, it is important to note that this process was not about oversimplifying the complexity of decision-making in megaprojects by merely categorising deviant tactics. Instead, the focus was on delving deeply into the subjective experiences of project team members and understanding their responses to deviations in practice. This book emphasised rich, detailed descriptions of these stories, capturing the nuances of the environment rather than relying on general statistical trends.

Chapter summary

Megaprojects, often cross-sectoral collaborations involving significant investments, technological innovation, and diverse stakeholders, are monumental not just in size but also in their societal and symbolic implications. Despite their potential to drive economic growth and societal development, the effectiveness of these projects is increasingly questioned due to their complex nature and frequent underperformance. This chapter delved into the inherent challenges of managing megaprojects. The decision-making environment of megaprojects transcends mere technical aspects, encompassing political, institutional, and psychosocial factors. This complexity demands a sophisticated approach that addresses the interconnected and often conflicting agendas. This chapter argued for a shift from idealised governance models to empirical explorations that consider the nuanced realities of power relations and decision-making in megaprojects. The next chapter provides a critical examination of the megaprojects discourse.

References

Altshuler, A., & Luberoff, D. (2003). *Megaprojects: The changing politics of urban public investment*. Washington, DC: Brookings Institution.

Biesenthal, C., Clegg, S.D., Mahalingam, A., & Sankaran, S. (2018). Applying institutional theories to managing megaprojects. *International Journal of Project Management*, 36(1), 43–54.

Bodankin, M., & Tziner, A. (2009). Constructive deviance, destructive deviance and personality: How do they interrelate? *Amfiteatru Economic*, 11(26), 549–564.

Boyce, G., & Davids, C. (2004). The dimensions of governmentality studies in accounting: Complementary and critical potentials, Fourth Asia Pacific Interdisciplinary Research in Accounting Conference, Singapore, July, APIRA and Singapore Management University.

Brady, T., & Davies, A. (2014). Managing structural and dynamic complexity: A tale of two projects. *Project Management Journal*, 45 (4), 21–38. DOI: 10.1002/pmj.21434

Bridge, A., In, S., & Rowlinson, S. (2015). *Models for engaging public-private partnerships in civil infrastructure projects: A case of 'having your cake and eating it too'?* In G. Macht, & C. Dossick (Eds.), *Proceedings of the 12th Engineering Project Organization Conference*. Engineering Project Organization Society (EPOS), United Kingdom, pp. 1–14.

Brookes, N. (2012). *Comparing the performanceof energy megaprojects across the EU*. Advanced Project Management for the Energy Industry, London, BIS.

Brunet, M. (2018). Governance-as-practice for major public infrastructure projects: A case of multilevel project governing. *International Journal of Project Management*, 37(2019), 283–297.

Brunet, M., & Aubry, M. (2018). The governance of major public infrastructure projects: The process of translation. *International Journal of Managing Projects in Business*, 11(1), 80–103.

Bruzelius, N., Flyvbjerg, B., & Rothengatter, W. (2002). Big decisions, big risks. Improving accountability in mega projects. *Transport Policy*, 9(2), 143–154.

Chan, P., & Jia, A. (2019). *Finding the micro in the mega: A multi-ethnographer study into the microdynamics in a mega construction project*. In C. Gorse, & C.J. Neilson (Eds.), *Proceeding of the 35th Annual ARCOM Conference*, 2–4 September 2019, Leeds, Association of Researchers in Construction Management, 1–10.

Cicmil, S., Williams, T., Thomas, J., & Hodgson, D. (2006). Rethinking project management: Researching the actuality of projects. *International Journal of Project Management*, 24(8), 675–686.

Clegg, S.R., Pitsis, T.S., Rura-Polley, T., & Marosszeky, M. (2002). Governmentality matters: Designing an alliance culture of inter-organizational collaboration for managing projects. *Organization Studies*, 23, 317–337.

Clegg, S.R., & Kreiner, K. (2013). *Power and politics in construction projects*. In N. Drouin, R. Müller, & S. Sankaran (Eds.), *Novel approaches to organizational project management research: Translational and transformational* (pp. 268–293). Copenhagen: CBS Press.

Clegg, S., Sankaran, S., Biesenthal, C., & Pollack, J. (2017). *Power and sensemaking in megaprojects*. In B. Flyvbjerg (Ed.), *The Oxford handbook of megaproject management* (pp. 238–258). Oxford: Oxford University Press.

Crawford, L., & Pollack, J. (2004). Hard and soft projects: A framework for analysis. *International Journal of Project Management*, 22, 645–653.

Dean, M. (2010). *Governmentality: Power and rule in modern society*. London: Sage Pubs.

DeBlois, M., Herazo-Cueto, B., Latunova, I., & Lizzeralde, G. (2011). Relationships between construction clients and participants of the building industry: Structures and mechanisms of coordination and communication. *Architectural Engineering and Design Management*, 6(1), 1–20.
de Bruijn, H., & Leijten, M. (2008). Management characteristics of mega-projects. In H. Priemus, B. Flyvbjerg, & B. van Wee (Eds.), *Decision making on megaprojects: Cost-benefit analysis, planning and innovation* (pp. 23–39). Cheltenham: Edward Elgar Publishing.
Denicol, J., Davies, A., & Krystallis, I. (2020). What are the causes and cures of poor megaproject performance? A systematic literature review and research agenda. *Project Management Journal*, 51(3), 328–345.
Fiori, C., & Kovaka, M. (2005). Defining megaprojects: Learning from construction at the edge of experience. ASCE Construction Research Congress: Broadening perspectives, April 5–7, California.
Flyvbjerg, B. (2014). What you should know about megaprojects and why: An overview. *Project Management Journal*, 45(2), 6–19.
Flyvbjerg, B., Bruzelius, N., & Rothengatter, W. (2003). *Megaprojects and risk: An anatomy of ambition*. Cambridge: Cambridge University Press.
Flyvbjerg, B. (2005). Design by deception: The politics of megaproject approval. *Harvard Design Magazine*, 22, 50–59.
Flyvbjerg, B. (2009). Survival of the unfittest: Why the worst infrastructure gets built—And what we can do about it. *Oxford Review of Economic Policy*, 25(3), 344–367.
Flyvbjerg, B., Skamris Holm, M.K., & Buhl, S.L. (2002). Underestimating costs in public works project: Error or lie? *Journal of American Planning Association*, 68(3), 279–295.
Foucault, M. (1986). Of other spaces. *Diacritics*, 16, 22–27.
Foucault, M. (1991). Questions of method. In G. Burchell, C. Gordon, & P. Miller (Eds.), *The Foucault effect: Studies in governmentality* (pp. 73–86). Chicago, IL: Chicago University Press.
Gil, N. (2009). Developing cooperative project-client relationships: How much to expect from relational contracts. *California Management Review*, 51(2), 144–169. DOI: 10.2307/41166484
Hallgren, M., & Wilson, T. (2007). Mini muddling: Learning from project plan deviations. *Journal of Workplace Learning*, 19(2), 92–107.
Jia, G., Yang, F., Wang, G., Hong, B., & You, R. (2011). A study of megaprojects from a perspective of social conflict theory. *International Journal of Project Management*, 29, 817–827.
Klein, G., & Aubry, M. (2017). From the editors: Introducing the special issue on megaprojects – Symbolic and sublime. *Project Management Journal*, 48(6), 3–4.
Li, H., & Guo, H. (2011). International Journal of Project Management special issue on "Complexities in managing mega construction projects". *International Journal of Project Management*, 29(7), 795–796.
Memon, S.A., Rowlinson, S., Sunindijo, R.Y., & Zahoor, H. (2021). Collaborative behavior in relational contracting projects in Hong Kong—A contractor's perspective. *Sustainability*, 13, 5375.
Merrow, E.W. (2011). *Industrial megaprojects: Concepts, strategies and practices for success*. Hoboken, NJ: Wiley.
Miller, R.J., & Lessard, D.R. (2000). *The strategic management of large engineering projects*. Cambridge: MIT Press.
Miller, R., & Lessard, D. (2008). Evolving strategy: Risk management and the shaping of mega projects. In H. Priemus, B. Flyvbjerg, & B. van Wee (Eds.), *Decision-making on*

mega projects: Cost-benefit analysis, planning and innovation (pp. 145–172). Cheltenham: Edward Elgar Publishing Limited.
Morris, P. (2009). Implementing strategy through project management: The importance of managing the project front-end. In T. Williams, K. Samset, & K.J. Sunnevåg (Eds.), *Making essential choices with scant information* (pp. 39–67). Basingstoke: Palgrave Macmillan.
Morris, P.W.G. (2013). *Reconstructing project management*. Hoboken, NJ: Wiley-Blackwell.
Morris, P., & Hough, G. (1987). *The anatomy of major projects*. London: Major Projects Association.
Ninan, J., Mahalingam, A., & Clegg, S. (2019). External stakeholder management strategies and resources in megaprojects: An organizational power perspective. *Project Management Journal*, 50(6), 625–640.
Novy, J., & Peters, D. (2012). Railway station megaprojects as public controversies: The case of Stuttgart 21. *Built Environment*, 38(1), 128–145.
Pinto, J.K. (1996). *Power and politics in project management*. Newtown Square, PA: Project Management Institute.
Pitsis, A., Clegg, S., Freeder, D., Sankaran, S., & Burdon, S. (2018). Megaprojects redefined – Complexity vs cost and social imperatives. *International Journal of Managing Projects in Business*, 11(1), 7–34.
Salet, W., Bertolini, L., & Giezen, M. (2012). Complexity and uncertainty: Problem or asset in decision making of mega infrastructure projects? *International Journal of Urban and Regional Research*, 37(6), 1984–2000.
Sanderson, J. (2012). Risk, uncertainty and governance in mega-projects: A critical discussion of alternative perspectives. *International Journal of Project Management*, 30, 432–443.
Soderlund, J., Sankaran, S., & Biesenthal, C. (2017). The past and present of megaprojects. *Project Management Journal*, 48(6), 5–16.
Steen, J., Ford, J., & Verreynee, M. (2017). Symbols, sublimes, solutions, and problems: A garbage can model of megaprojects. *Project Management Journal*, 48(6), 117–131.
van Marrewijk, A. (Ed.) (2015). *Inside megaprojects: Understanding cultural practices in project management*. Copenhagen: CBS Press.
van Marrewijk, A., Clegg, S.R., Pitsis, T., & Veenswijk, M. (2008). Managing public-private megaprojects: Paradoxes, complexity and project design. *International Journal of Project Management*, 26, 591–600.
Volden, G.H., & Andersen, B. (2018). The hierarchy of public project governance frameworks: An empirical study of principles and practices in Norwegian ministries and agencies. *International Journal of Managing Projects in Business*, 11(1), 174–197.
Wachs, M. (1989). When planners lie with numbers. *Journal of American Planning Association*, 55(4), 476–479.
Warren, D. (2003). Constructive and destructive deviance in organizations. *The Academy of Management Review*, 28(4), 622–632.
Williams, T., & Samset, K. (2010). Issues in front-end decision making on projects. *Project Management Journal*, 41(2), 38–49.
Winch, G.M., & Leiringer, R. (2016). Owner project capabilities for infrastructure development: A review and development of the "Strong Owner" concept. *International Journal of Project Management*, 34(2), 271–281.
Winter, M., Smith, C., Cooke-Davies, T., & Cicmil, S. (2006). The importance of 'process' in rethinking project management: The story of a UK government-funded research network. *International Journal of Project Management*, 24, 650–662.

2 Megaproject decision-making

Current challenges

Defining megaprojects

The term *megaproject* is relatively new to academia despite large projects being built for centuries (Soderlund et al., 2017). The first known use of the word was documented in 1976 by the Merriam-Webster Dictionary. The term has been used to describe projects that push the limits of construction size, scope, scale, and experience. Megaprojects are "large-scale, complex ventures that typically cost US$1 billion or more, take many years to develop and build, involve multiple public and private stakeholders, are transformational, and impact millions of people" (Flyvbjerg, 2017, p. 2). Megaprojects are commonly initiated by government bodies or public organisations, which possess the requisite resources, motivation, and time to commit to these substantial endeavours (Shenhar & Holzmann, 2017).

Numerical thresholds (e.g. US$1 billion) have been used to define megaprojects (Flyvbjerg et al., 2003; van Marrewijk et al., 2008). While a commonly accepted benchmark for classifying megaprojects is a cost exceeding 100 million euros (Hu et al., 2015), the criteria that elevate a project to "mega" status extend beyond just financial magnitude. Fiori and Kovaka (2005) describe a megaproject, or a collection of such projects, as entities marked by amplified costs, heightened complexity, and escalated risks. These projects are also distinguished by their ambitious objectives and prominent visibility. This combination presents a considerable challenge to all stakeholders involved, exerts a substantial influence on the community, and stretches the boundaries of existing construction expertise (Fiori & Kovaka, 2005).

It is noteworthy that while many megaprojects are infrastructure-related, their scope is not limited to this category. Megaprojects can encompass a diverse array of ventures, including major events, theme parks, museums, stadiums, hospitals, and large-scale signature architecture (Desjardins, 2017; Altshuler & Luberoff, 2003). Moreover, there is a growing anticipation that future megaprojects will extend to initiatives aimed at addressing global challenges like disease control and extreme weather conditions. This evolution signifies not only an expansion in the scale of these projects but also a broadening in the variety of undertakings classified as megaprojects.

DOI: 10.1201/9781003377481-2

Box 2.1 Megaproject characteristics.

Some key characteristics of megaprojects include (Fiori & Kovaka, 2005; van Marrewijk et al., 2008; Biesenthal et al., 2018; Soderlund et al, 2017):

- *Colossal*: in size and scope with many of these projects spanning across decades, usually highly visible once construction begins.
- *Captivating and idealistic*: because of size, engineering achievements, and aesthetic design.
- *Costly*: costs are often enormous and tend to increase over the life of the project.
- *Controversial, highly visible, and politically sensitive*: funding, mitigation packages, impacts on third parties.
- *Complex*: particularly high number of risks and uncertainty in terms of design, funding, and construction.
- *Control issues*: involvement of a large number of partners and stakeholders which results in difficulties in decision-making, management/operations, and funding.
- *Transformational*: their reach and economic, social, and environmental impacts are broad and lasting.

Each of these characteristics is overly magnified beyond the level of "standard", thereby resulting in a megaproject (Fiori & Kovaka, 2005). Although these characteristics are largely relevant to any complex or large project, a key differentiating factor of megaprojects is their wide reach and broad impact on society and the environment (Biesenthal et al., 2018).

Megaprojects have been termed "trait making" (Hirschman, 1995), as they are often designed with the intention to transform societal structures (Flyvbjerg, 2014). Although many globally known megaprojects exhibit universally accepted characteristics of "mega-ness", the perception of what constitutes a megaproject is significantly influenced by the context of the project. This includes the perspectives of involved stakeholders, the impacted community, and the characteristics of the host environment (Fiori & Kovaka, 2005).

This contextual nature means that a project deemed "mega" in one setting might not be perceived similarly in another. Certain projects are considered megaprojects in specific contexts because they challenge local norms or exist at the limits of what the stakeholders and community are familiar with or understand (Fiori & Kovaka, 2005). Consequently, what might be categorised as a large-scale project in an environment with relatively straightforward political and organisational relations could be viewed as a megaproject in a different setting due to its complexity, novelty, or impact (Pitsis et al., 2018).

Challenges in managing megaprojects

Megaprojects are often characterised as a distinct and uniquely challenging class of projects. Flyvbjerg (2014) describes them as "a completely different breed of project", necessitating a distinct approach to management. Their complexity and scale have led to various metaphorical descriptions, such as being the "wild beast" of the project world (Zidane et al., 2013), "some kind of uncontrollable species" (Pitsis et al., 2018), and "unruly trolls" (Klakegg et al., 2016). These descriptors capture the notoriously difficult nature of controlling and managing megaprojects.

Their colossal size, scope, and costs are just surface-level challenges. Deeper intricacies arise from the complex political dynamics involved in securing funding, managing, and governing the intricate web of social and organisational relationships (Pitsis et al., 2018; Fiori & Kovaka, 2005). As Giezen et al. (2015) explain, the planning and decision-making process in megaprojects can be likened to navigating a vast ocean, replete with its own set of challenges such as storms, lulls, whirlpools, and sharks.

The chronic underperformance of megaprojects is a well-documented issue, extensively discussed in both industry and academic circles. This challenge has even led to the creation of specific jargon within the field. Notably, Flyvbjerg (2014) coined the term the "Iron Law" of megaprojects, encapsulating the tendency of these projects to be "over budget over time, over and over again".

Despite their history of underperformance, there is ongoing and robust investment in megaprojects globally. This persistent trend of escalating investment in megaprojects, despite their frequent failures, has been termed the "megaproject paradox" (Flyvbjerg et al., 2003). It reflects the intriguing contradiction between the allure and the actual performance of these massive projects.

The fascination and optimism surrounding the development of megaprojects are undeniable. However, they bring with them a host of uncertainties and risks that are not typically encountered in smaller, everyday projects (Javernick-Will & Levitt, 2009).

Megaprojects represent far more than merely scaled-up versions of smaller, conventional projects (Flyvbjerg, 2014). Their unique characteristics place them "at the edge of technological and institutional complexity", imposing exceptional demands on the management capabilities required for their successful execution (Soderlund et al. 2017; p. 6). These projects operate within an environment of intense pressures, necessitating rapid development of relationships and coordination at levels rarely seen in typical projects.

This heightened level of complexity and challenge in megaprojects calls for a specific type of project manager. As Flyvbjerg (2014, p. 6) emphasises, managing megaprojects requires practitioners who possess not just general project management skills but deep, specialised experience in the specific domain of the project. The analogy he uses is striking: "And just like you wouldn't want someone with just a driver's license to fly a jumbo jet, you wouldn't want conventional project managers to manage megaprojects".

Charting the megaproject discourse

Megaproject research, while primarily rooted in the project management discipline and often considered a subset or specialisation within it, has garnered substantial interest across a diverse array of academic fields. This review focuses on tracing and classifying the academic literature relating to the megaproject phenomenon. The body of work on this subject can be broadly categorised into four key disciplinary areas of influence:

- Architecture, engineering, and construction (AEC)
- Project management
- Organisation and management
- Urban development and planning

Figure 2.1 serves as a visual representation, mapping out significant works and developments in the four disciplinary areas related to megaproject research over the three decades from 1990 to 2020. This diagram clearly illustrates the burgeoning interest in megaprojects attracting attention from both the research community and policymakers, particularly in the last decade. The latter part of the timeline, particularly in the last ten years, shows a significant increase in academic output and interest in megaprojects, as evidenced by the publication of several special issues in various scholarly journals. This uptick in research activity demonstrates the growing recognition of the importance and complexity of megaprojects, as well as the need for continued study and exploration in this field.

The megaproject boom, particularly around 2008, catalysed a heightened focus on understanding and improving the management of such large-scale projects. Despite this growing interest, management research specifically dedicated to megaprojects is a relatively recent development.

This area of research is still in its nascent stages, with seminal works and "classic texts" in the field having emerged only in the past decade or so (Flyvbjerg & Turner, 2017). Notably, empirical studies that delve into the realities of megaprojects have been somewhat limited, especially when compared to the volume of research conducted on other types of projects. This scarcity of empirical data has left many aspects of megaproject research relatively unexplored.

Architecture, engineering, and construction

Megaprojects often grapple with the challenges of substantial cost overruns and high public visibility. Given these high-stakes conditions and the considerable risks associated with potential failures, there is an increased emphasis on the effective management of such projects.

In response to these challenges, a substantial portion of research and professional practice in the AEC sector has been dedicated to developing and refining methods for risk management. This research focuses on tools and techniques for identifying, assessing, and managing the risks and uncertainties inherent in megaprojects.

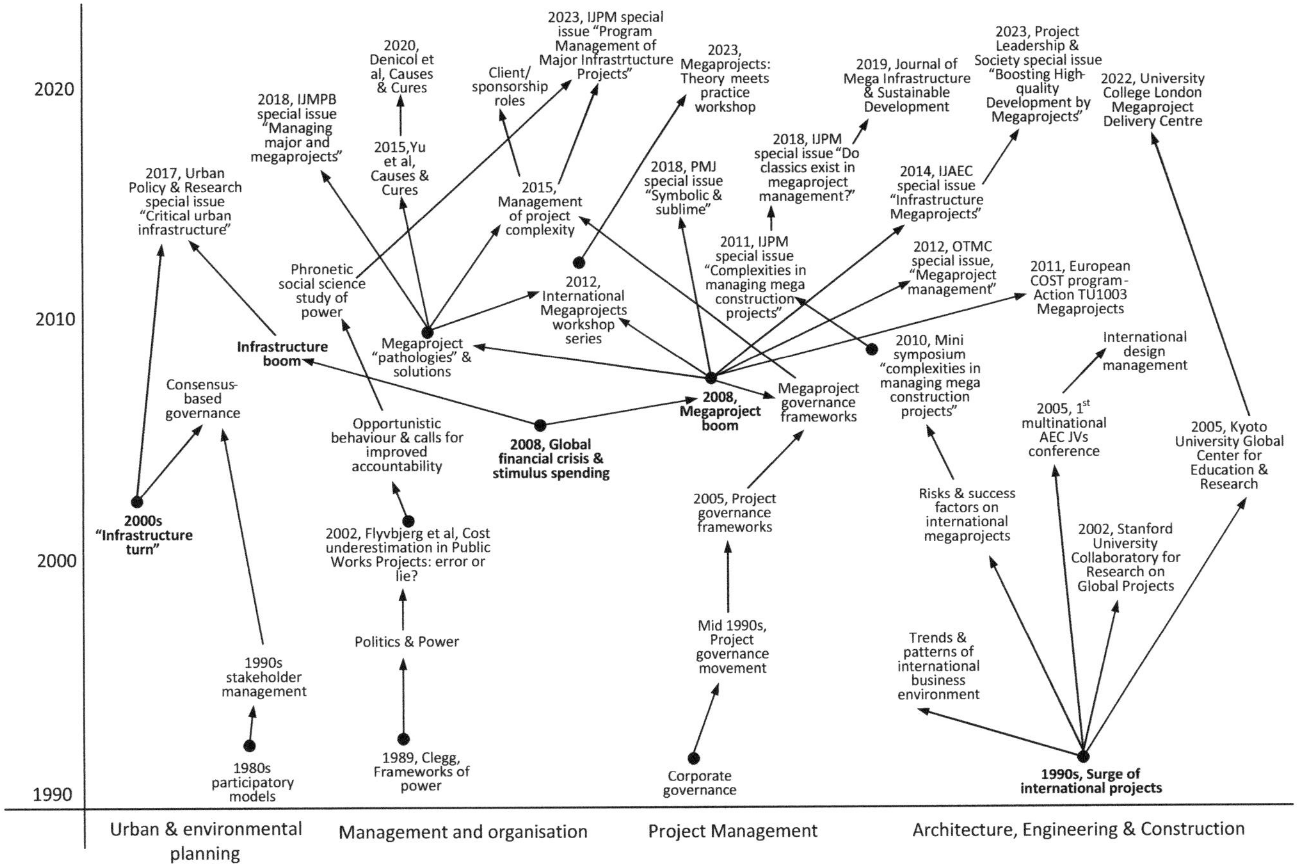

Figure 2.1 Key works and developments in megaproject research from 1990 to 2020.

Global megaprojects in the AEC sector

The 1990s marked a pivotal era with a notable surge in large-scale global megaprojects, particularly drawing attention to their management within the AEC sector. This period sets in motion significant advancements in the international discourse on megaprojects. A central theme of these studies has been to provide comprehensive overviews of the international business environment. This encompasses examining broad-based trends and patterns that are influential in shaping the landscape of global construction (Ofori, 1994; Bon & Crosthwaite, 2000; Mawhinney, 2001; Zhang & London, 2012). Key areas of focus in these studies include the volume of international construction, key markets, competitive forces, trade patterns, and major players in the global construction market.

In 2005, a significant milestone in the study of global AEC projects was marked by the first conference on multinational joint ventures in construction. Since then, annual international conferences have continued, with key insights published in works like "*Joint Ventures in Construction*" by Kobayashi et al. (2009), offering an international perspective on joint ventures and partnering in construction projects.

Additionally, research centres like the Collaboratory for Research on Global Projects (CRGP) at Stanford University, the Global Centre for Education and Research on Human Security Engineering for Asian Megacities at Kyoto University, and the Megaproject Delivery Centre at the University College London have been established. These centres serve as hubs for studies, collaborations, and debates on the management of global AEC megaprojects.

Discussions within these conferences and research centres have largely focused on exploring the characteristics of successful international firms in the AEC sector. Moreover, broader aspects of internationalisation within design management have been addressed, including topics like long-term sustainability in international markets (London et al., 2005; Grilo et al., 2007) and the evolving sophistication of international clients (Ding & Poon, 2009).

The exploration of internationalisation in the AEC sector includes applying theoretical models from international business to understand the performance and strategic behaviours of firms engaged in international construction. Notable among these are Porter's diamond framework (1998) and Dunning's eclectic paradigm (2000), which have been frequently used to analyse the sector (e.g. Ofori, 1994; Oz, 2001; Abdul-Aziz et al., 2008). These studies typically focus on the competitive dynamics of the international construction market. This approach helps in understanding the factors that drive success in international construction and guides firms in making strategic decisions to enhance their global performance.

There has also been a substantial trend toward revealing risks and challenges that affect AEC firms engaged in global megaprojects. Some of these past studies have identified enablers or success factors specifically related to categories of international work, including:

- Project delivery methods (Kumaraswamy & Morris, 1998; Shou Qing et al., 1999; Chan & Tse, 2003; Kobayashi et al., 2009; Pena-Mora & Tamaki, 2001)

- Social, cultural, and normative or institutional differences (Mahalingam & Levitt, 2007; London & Siva, 2010; Javernick-Will & Scott, 2011; Chi & Javernick-Will, 2011)

Box 2.2 Case study – factors influencing the performance of firms in international markets.

The AEC research community has increasingly adopted empirical case studies as a method to uncover factors influencing project performance. An example is the work of Javernick-Will and Scott (2011), who developed the "three pillars framework" based on an analysis of 14 case studies across three types of international firms. This study made significant strides in identifying and categorising the types of institutional knowledge crucial for firms operating in international markets. It classified this knowledge into three categories: normative, regulative, and cultural. Among these, normative knowledge – which related to the professional practices that need to be adapted and improvised in use – was highlighted as the most vital for firms to acquire.

Megaproject management: risks and success factors

Alongside developments in global megaprojects research, there has been a body of work undertaken specifically to investigate the complexities and challenges related to the management of megaprojects in the AEC sector. A key event contributing to this discourse was a mini symposium titled "Complexities in managing mega construction projects" held in Hong Kong in December 2010. Building on the insights from this symposium, in 2011, the *International Journal of Project Management* published a special issue, compiling a total of eight papers, each reflecting on contemporary and emerging themes in megaproject research. Further contributing to the scholarly conversation, the "*Organisation Technology and Management in Construction*" journal dedicated a special edition to "megaproject management" in 2012. This edition featured six contributions, each exploring different facets of managing megaprojects.

This body of work has been important in establishing and identifying several key areas for development in megaproject management research in the AEC sector including:

- Stakeholder management (Yang et al., 2011)
- Risk assessment and management (Bridge et al., 2015; Boateng et al., 2012)
- Project delivery models such as PPPs
- Management of project complexity (Antoniadiis et al., 2011; London & Siva, 2010)

The predominant focus of this body of work has been on the development of models or tools grounded in rational thinking and analytical approaches. This emphasis is reflected in works that propose systematic frameworks and methodologies to enhance the management of megaprojects. For example, see Boateng et al. (2012) system dynamics model for risk management in megaprojects.

During this period, the European Cooperation in Science and Technology (COST) programme played a key role in advancing megaproject research. It established a research network comprising over 80 multi-disciplinary researchers from 24 European countries. The aim was to deepen the understanding of the design and delivery of megaprojects (Brookes et al., 2014).

Box 2.3 Rethinking megaproject management – a critique of traditional approaches and the shift towards holistic perspectives.

A significant contribution to the discourse came in 2014 with the "*International Journal of Architecture Engineering and Construction*" publishing a special issue titled "Infrastructure megaprojects". This special issue critically examined traditional management approaches and tools, questioning their adequacy in addressing complexities and challenges inherent in megaprojects. It highlighted the necessity for a new philosophical perspective and a set of decision-making instruments tailored to improve outcomes in managing megaprojects.

A key point of critique in this issue was the past research emphasis on project efficiency over effectiveness. This focus relied heavily on the precise application of tools and techniques (Kusuma, 2014). However, it often overlooked the intricate social–cultural systems within which megaproject temporary multi-organisations (TMOs) operate. This critique marked a significant shift from the prevailing trend of developing analytical frameworks and processes primarily aimed at enhancing technical and operational accuracy and efficiency. It highlighted the need to consider broader, more complex organisational and socio-cultural factors in megaproject management beyond mere technical efficiency.

Project management

Megaprojects are distinguished by their high level of technical complexity, which can be a primary source of the challenges they face. Many issues encountered in these projects can be linked to shortcomings in the analytical procedures used during critical decision-making stages of the project lifecycle. Key phases such as the initial decision to proceed, scope definition, and the ongoing interaction between analysts and decision-makers are pivotal moments where these deficiencies can have significant impacts (Samset et al., 2006).

Reflecting this perspective, a prevailing approach within the project management research community has been to attribute poor performance in megaprojects to a range of technical challenges encountered during project delivery (Siemiatycki, 2016). This viewpoint posits that the difficulties in executing megaprojects successfully are largely due to technical complexities, including issues related to design, engineering, construction, and integration of advanced technologies.

Atkinson et al. (2006) point out that a fundamental goal of project management is to significantly reduce uncertainty about achieving specific objectives. Reflecting this goal, a considerable focus of past project management research has been on developing analytical tools, processes, and procedures. These are designed to enhance the accuracy of data and information, thereby facilitating improved decision-making, effective stakeholder management, and efficient supply chain integration.

In line with these objectives, there has been substantial work in advancing forecasting techniques and methodologies. The aim here is to refine project planning, governance, and procurement processes to minimise risks and uncertainties associated with project execution (e.g., Priemus, 2008).

The project governance movement

Governance theory, which originated in political science (Nachmias & Greer, 1982), has since expanded its application across various disciplines, including in the construction sector. Over the past two decades, the concepts of governance have become increasingly prominent in project management literature (e.g. Samset et al., 2006; Klakegg et al., 2008; Miller & Lessard, 2008), indicating a shift in focus from solely technical and operational efficiencies to a broader interest in the dynamics of stakeholder interactions and their impact on project outcomes. In the context of construction, the application of governance theory can be categorised into two primary components, including corporate and project governance.

Corporate governance in the context of construction and project management involves the organisation and coordination of interactions among various actors, firms, and institutions. It emphasises transparency and accountability in these interactions (Jacoby, 2005; Sanderson, 2012). Corporate governance is "a system that shapes who makes investment decisions in corporations, what types of investments they make, and how returns from investments are distributed" (O'Sullivan, 2003, p. 24). This system plays a crucial role in guiding the strategic direction and operational practices of corporations.

Project governance focuses on facilitating efficient and effective decision-making specific to individual projects (Garland, 1997). It aims to establish a structured framework that encompasses three critical areas: the project objectives, the means to achieve those objectives, and the means to monitor project performance (Turner et al., 2009). In essence, project governance is the arena in which day-to-day project activities take place (Patel, 2007). It includes a system that enforces "appropriate checks and balances that enables transparency, accountability and defined roles in the projects" (Volden & Andersen, 2018).

The dimensions of corporate and project governance, while distinct in their focus and scope, are intrinsically interconnected. Efficient and effective governance at both levels is crucial for aligning project outcomes with broader corporate strategic objectives within an overarching corporate governance framework (Müller, 2009). This interrelation emphasises the importance of cohesive governance practices that bridge the gap between project-level execution and corporate-level strategy. Despite the variability in the description and application of governance across different contexts, there are several core elements commonly shared in its definition. Governance, irrespective of the setting, typically encompasses aspects such as rules, structure, enabling activity or change, and reaching goals.

Box 2.4 The project governance movement.

In the field of project management, the concept of governance initially received little attention in relation to specific projects until the mid-1990s. Post this period, there has been a marked shift, with a growing interest in exploring the governance of projects, especially in the context of large, complex projects or megaprojects (Miller & Hobbs, 2005). This burgeoning interest has been particularly pronounced in the study of the application of governance principles at the project level.

This trend has been characterised as a "project governance movement" (Patel, 2007), indicating a shift towards a more focused and deliberate examination of the processes and procedures that underpin effective project governance frameworks, which gained significant momentum around 2005. The "project governance movement" reflects a broader recognition of the importance of structured, systematic governance in achieving project success, particularly in the management of megaprojects.

Project governance frameworks

Project governance frameworks typically consist of structures, processes, decision-making tools, and systems meticulously designed to ensure optimal quality from the outset. They effectively manage risks, ensure adherence to agreed objectives, and address project issues. Additionally, these frameworks establish standards for reviewing the quality of key project milestones (Turner & Keegan, 2001; Crawford & Cooke-Davies, 2009; Samset et al., 2006, PMI, 2013). A governance framework is concisely defined as: "an organised structure established as authoritative within the institution, comprising processes and rules established to ensure projects meet their purpose" (Klakegg et al., 2008, p. 30).

Box 2.5 Project governance frameworks.

The critical role of governance frameworks in shaping the planning and management of projects is widely acknowledged (Klakegg et al., 2008, APM, 2002; Miller & Lessard, 2000). Recognising their significance, governments globally have implemented various governance frameworks for large public investment projects, with some being obligatory. Some examples include:

- The UK Office of Government Commerce (OGC) and the UK Ministry of Defense (MoD)
- The Quebec Governance Framework (QGC)
- The PRINCE2 methodology
- The Norwegian Ministry of Finance

Additionally, prominent organisations such as the Association of Project Management (APM) have established comprehensive standards and guidelines. APM, for instance, has introduced 13 principles for the governance of project management (Volden & Andersen, 2018).

The academic community has also contributed significantly to the discourse on project governance. A diverse range of publications have merged, offering various perspectives and insights into the subject. Noteworthy contributions include works by Bruzelius et al. (2002), Sha (2016), Klakegg et al. (2008), Dewulf and Garvin (2020), and Miller and Lessard (2008).

Box 2.6 Case study – a systems analysis methodology to project governance.

In their examination of decision-making processes in the early stages of megaprojects, Priemus et al. (2008) observed a common pattern where a preferred solution often emerges early in the project lifecycle. This preferred solution, typically aligning with the interests of initiators or sponsors, tends to be advanced without a thorough analysis of alternative options. Due to the "lock-in" nature of these large-scale projects, where significant resources and commitments are made early on, alternative solutions proposed at later stages are frequently deemed infeasible or too late to be effectively implemented.

To counteract this issue, Priemus et al. (2008) advocate for the adoption of the systems analysis methodology. The methodology commences with an in-depth analysis of the problem. Following the problem analysis, a range of alternative solutions is generated. The systems analysis methodology encourages decision-making based on a comprehensive understanding of both the problem and the possible solutions.

Box 2.7 Case study – the evolutionary systems approach to project governance.

To address the argument that the uniqueness of each project makes generalisation impossible, Miller and Lessard (2008) conducted an extensive study aimed at identifying factors contributing to the success or failure of megaprojects. Their research involved an in-depth assessment of 60 large engineering projects spanning Europe, North and South America, and Asia.

Drawing insights from this diverse array of projects, Miller and Lessard proposed a governance framework adopting an evolutionary systems perspective, recognising that while projects may be unique in many respects, there are common underlying principles that can guide their successful management.

The evolutionary perspective represents a significant shift from traditional project management approaches. Rather than measuring project performance solely based on how outcomes align with initial goals, this perspective emphasises the dynamic processes of shaping, countermoves, and risk management throughout the project lifecycle. Projects are viewed as a series of interdependent episodes where decisions are made through a process of creating and shaping. Decision-making is seen through the lens of real options, involving a sequence of creating, exercising, or abandoning choices based on the evolving project context. This strategy is crucial in managing uncertainty, preserving value, and enabling a swift move to completion once the direction is clear. The critical influence of resistors, or those who sense potential value destruction, in shaping project episodes was raised by Biesenthal et al. (2018). These stakeholders play a vital role in challenging decisions and directions.

The governance frameworks proposed by Miller and Lessard (2008) and Priemus et al. (2008) offer a good representation of the general approach undertaken by the project management research community in efforts to enhance megaproject performance. However, the governance of megaprojects continues to pose significant challenges (Sha, 2016; Samset & Volden, 2015). Despite the wealth of guides, advice, and information available, practical guidance on governance often remains limited primarily focused on traditional control-based management approaches embodied in the plan-do-check-act (PDCA) cycle (Sha, 2016). Key observations include:

1 *Narrow focus of governance frameworks*: Many project governance frameworks primarily concentrate on classical governance means, addressing technical and operational tasks at a tactical level (Klakegg et al., 2008). This often overlooks the broader aspects of governance, such as steering multiple agencies and stakeholders.

2 *Human aspect of governance*: While Miller and Lessard's (2008) project shaping model makes a departure from mechanistic project management approaches, it has been critiqued for not fully capturing the human side of governance, particularly the influence of various stakeholders on decision-making practices.
3 *Front-end focus*: A considerable amount of project governance research has been directed towards improving front-end design and planning of megaprojects. However, there has been less emphasis on decision-making processes and problem-solving from construction through to completion (Steen et al., 2017; Olaniran et al., 2016).

The characterisation of megaprojects as entities marked by interdependence and uncertainty challenges the efficacy of traditional project management approaches and governance frameworks that typically view project development as an orderly, linear process. Scholars such as Hagan et al. (2012), Remington and Pollack (2007), and Dewulf et al. (2016) argue that these conventional models fail to accurately reflect the complex realities of megaprojects. First, traditional approaches often presume that megaproject development can be methodically organised, planned, and controlled in a top-down manner, which may not adequately account for the dynamic and unpredictable nature of these projects. Second, an over-focus on operational planning and control can lead to the oversight of significant strategic issues (Sanderson, 2012). Third, the inherent complexity and uncertainty of megaprojects require a flexible and adaptive approach to management and governance. Traditional models, with their structured and linear frameworks, often lack the necessary agility to respond effectively to unforeseen changes and challenges.

Addressing project complexity

Project management researchers have increasingly turned to a socio-cultural perspective to understand and address the significant performance problems associated with megaprojects. This approach acknowledges that project environments are not just technical constructs but are also shaped by complex social dynamics and interactions (van Marrewijk et al., 2008; Pitsis et al., 2018; Geraldi, 2008; Brady & Davies, 2014; Morris & Hough, 1987). Therefore, there is a need to recognise these complexities that co-exist and operate within megaprojects in terms of ambiguity, conflicts, and power relations.

Megaprojects are defined by a culture that is inherently ambiguous, marked by indistinct boundaries and a dual nature involving both objects and actors actively shaping the project's existence (van Marrewijk et al., 2008). These projects typically lack a singular locus of control, instead featuring numerous collaborators. The fragmentation perspective on organisational culture, which emphasises the roles of ambiguity, power dynamics, and situational factors, proves more enlightening for understanding the daily practices in megaprojects (van Marrewijk et al., 2008).

In the socio-cultural context of megaprojects, the interpretation of project contracts and documents varies significantly among stakeholders, each bringing their own worldviews to the table. This diversity in perspectives often leads to ambiguities in project coordination (Clegg & Kreiner, 2013). Recognising these diverse viewpoints, resolutions are often sought through negotiations and debates among multiple stakeholders (Pitsis et al., 2018). The process of negotiation in such complex environments can give rise to what van Marrewijk et al. (2008) describe as "language games" and "contested action in complex inter-organisational and professional disputes". In this intricate setting, project team members are often likened to "skilled game players" (Pitsis et al., 2018). They are adept at navigating the complex landscape of stakeholder interactions, leveraging opportunities, and steering the project through a web of competing interests and viewpoints.

In September 2013, a significant development in the field of megaproject research occurred with the inauguration of a workshop titled "Megaprojects: theory meets practice" in the United Kingdom. This workshop was designed to serve as a platform for fostering theoretically robust and empirically well-grounded research on megaprojects. Since its inception, this annual workshop has been hosted in various global locations, each year focusing on different themes relevant to megaprojects, such as artificial intelligence, challenges in emerging markets, and the interplay of power, politics, and value creation in megaproject contexts. These forums have highlighted the importance of developing a solid theoretical foundation to better understand the complexities and realities of megaprojects. Discussions at these gatherings have emphasised the need for sophisticated models and frameworks that can effectively address the multi-faceted nature of megaprojects.

Box 2.8 Fuzzy multi-attribute decision-making (FMADM).

The endeavour to establish a shared project culture and minimise ambiguity in megaprojects often employs systems and methods that align with those discussed in project governance frameworks. A notable approach is the integration of risk management tools and methods that leverage "fuzzy" reasoning and multi-attribute decision-making (MADM), collectively referred to as fuzzy MADM (FMADM). This approach is designed to address the inherent uncertainties in the decision-making environment of megaprojects, including vagueness, ambiguity, and the subjectivity of expert judgments.

Key applications of FMADM in megaproject risk assessment include:

- Modified fuzzy analytical hierarchy process (AHP) (Zeng et al., 2007)
- Fuzzy multi-attribute group decision-making approach (Makui et al., 2013)
- Fuzzy Technique for Order Preference by Similarity to Ideal Solution (FTOPSIS) (Mousavi et al., 2013)
- Fuzzy compromise approach (Ebrahimnejad and Tavana, 2014)

Box 2.9 Rational models and tools.

The development of models and tools in megaproject management has seen the incorporation of game theory to systematically identify risks and prepare for uncertain scenarios (e.g. Khallaf et al., 2018). Game theory offers a structured approach to understanding interactions and decision-making processes under conditions of uncertainty, making it particularly valuable in the context of megaprojects. Researchers have also focused on specific areas within megaproject risk analysis and management, including:

- Risk identification (Santoso et al., 2003; Busby & Hughes, 2004)
- risk measurement
- risk control (Flyvbjerg, 2006)

While these contributions in areas of fuzzy reasoning, multi-attribute decision-making, and risk analysis and management are valuable, gaps remain in understanding how project governance practices can effectively cope with uncertainty and how project teams respond to unforeseen events or conflicts (Dewulf & Garvin, 2020). The common "closed" approach to planning megaprojects, often encapsulated in the KISS ("keep it simple stupid!") mantra, seeks to minimise complexity and uncertainty. However, this approach might be counterproductive, as it can reduce the resilience and potential added value of these projects (Giezen et al., 2015).

Leadership and the client/sponsor's role

Leadership plays a key role in project-based industries like construction. The construction industry, characterised by its project and team-based nature, often encounters ambiguities, conflicts, and power dynamics that can significantly impact project outcomes. Effective leadership is essential in navigating these complexities and fostering a collaborative and productive work environment. Creating the conditions for effective team formation and collaboration is crucial, and good project leadership is one of the primary drivers of this process.

The role of the client in project management, especially in the context of megaprojects, has gained significant recognition in recent literature (Brady & Davies, 2014; Crawford et al., 2008). Contextual factors, such as top management support and leadership, are increasingly seen as vital to project success.

While project-based management and the competence of practitioners are important, evidence suggests that project success is not solely within the control of the project manager and team. The client's leadership, particularly in governance and support, is equally crucial (Crawford et al., 2008; Briscoe et al., 2004). Crawford et al. (2008) identified key competencies for client champions or project sponsors to enhance their effectiveness. These include good communication skills, commitment, a strategic position within the organisation, and availability to engage with the project team.

Box 2.10 Case study – the importance of client leadership in project management.

The study by Rose and Manley (2010) on four large Australian building projects commissioned by government clients provides valuable insights into the use of financial incentives to improve project outcomes. Their research highlights several critical aspects of client involvement and leadership in project management. The study found that designing effective risk/reward systems to motivate stakeholders requires a delicate balance. These systems need to be structured in a way that aligns the motivations of all stakeholders, particularly in projects characterised by high interdependence and contractual fragmentation across multi-organisational teams.

Clients play a crucial role in setting the tone and culture of a project. This includes implementing strategies such as equitable contract risk allocation, involving contractors early in the design process, focusing on value-driven tender selection, organising relationship-building workshops, and providing opportunities for future work. Successful financial incentives are dependent on their ability to foster a sense of unified motivation among all project stakeholders, ensuring that everyone is working towards common goals. With an adequate understanding of the stakeholders' context, clients can assume a leadership role that facilitates positive cultural change within the project environment.

The findings by Rose and Manley (2010) are supported by Brady and Davies (2014), who identified strong client leadership and capabilities as key success factors in their case studies of Heathrow Terminal 5 and the London 2012 Olympic Park. Clients have the potential to significantly influence project outcomes, not only through financial incentives but also by fostering a collaborative and motivated project environment.

Gil (2009) delved into the challenges clients encounter when striving to foster cooperative behaviour among project stakeholders, particularly through the implementation of relational contracting strategies. This approach in contracting, while aimed at promoting collaboration, requires careful and strategic implementation to be effective. The research reveals several important implications for management. One critical finding is the necessity of establishing suitable incentives. These incentives are essential for fostering cooperative behaviour among employees. Additionally, introducing a sense of competition among suppliers is beneficial. This competition helps in preventing any decline in their performance levels.

Another significant aspect is the adoption of a relational strategy. However, as Gil (2009) emphasises, it's not just the choice of this strategy that matters. Equally important is the way it is executed in practice.

Currently, there is a noticeable gap in understanding how clients act as change agents in megaprojects. Specifically, the process of creating an ideal environment that leads, supports, or encourages innovation is not well-documented. An area that demands more attention is the leadership role of the client, particularly how they influence power dynamics within projects.

Leadership research in construction has been predominantly concentrated on managerial aspects (Chan & Cooper, 2007). This narrow focus overlooks the wider sociological, psychological, and political dimensions of leadership. Additionally, previous studies have primarily focused on the role of built environment professionals as leaders, largely neglecting the client's role in leadership.

Explaining megaproject paradoxes

Megaprojects present a unique set of challenges, often manifested as multiple paradoxes. To gain a deeper understanding of these underlying paradoxes, several theoretical perspectives have been proposed. These perspectives help to explore the tension between logical, rational planning, and the often-unpredictable political realities that influence decision-making in megaprojects. Understanding these tensions is crucial for effectively navigating the challenges inherent in managing such large-scale and complex projects.

Box 2.11 Theoretical perspectives of project governance frameworks.

In their comparative study of three governance frameworks in the United Kingdom and Norway, Klakegg et al. (2008) identified notable differences between the frameworks, despite their shared objectives. To understand these differences and their implications, Klakegg et al. (2008) employed a diverse array of theoretical perspectives, drawing from political science and economics.

Their analysis included economic-rational (Eisenhardt, 1989; Thatcher, 2005; Christensen, 2007; Clarke, 2004; Jones et al., 1997; Williamson, 1979; Peltzman, 1998), instrumental-structural, cultural-institutional (March & Olsen, 1989), environmental (Meyer & Rowan, 1977), and "garbage-can" (March & Olsen, 1989). This multi-theoretical approach is particularly essential in understanding megaproject governance, which, due to its complexity, cannot be fully comprehended through a single theoretical lens.

The study by Klakegg et al. (2008) highlights the instrumental-structural perspective as particularly effective in explaining the essence of governance frameworks. This perspective emphasises the importance of formal structures in shaping thoughts and actions while acknowledging the limitations of rationality.

In addition to the instrumental-structural perspective, other theoretical views were found useful in elucidating the variances across governance frameworks. For instance, principal agent theory was instrumental in understanding the distinct control measures implemented in each framework. These measures aim to minimise information asymmetry and curb opportunistic behaviour.

The cultural-institutional perspective also played a vital role, explaining that differences arise from what is considered appropriate within each country and sector. This perspective focuses on the norms and values that influence governance practices.

A significant finding of the study is that for governance frameworks to be effective they need to be firmly rooted in both political and administrative contexts. This grounding ensures their relevance and practicality.

While Klakegg et al. (2008)'s study concentrated on governance frameworks, other scholars have explored project complexity in megaprojects, offering alternative theories grounded in sociological perspectives. These perspectives delve into the intricate, high-pressure, and often contentious decision-making environments characteristic of megaprojects (Steen et al., 2017).

Complexity theory has been increasingly applied to understand project complexity (Baccarini, 1996; Remington & Pollack, 2007; Brockmann & Girmscheid, 2007; Geraldi, 2008). This approach focuses on the intricate and often unpredictable nature of projects. Contingency theory examines how project management must adapt to varying circumstances (Maylor et al., 2011). Systems theory views projects as complex systems comprising interrelated components (Hobday, 1998; Shenhar & Dvir, 2007).

Brady and Davies (2014) built on these concepts and developed a framework that distinguishes two main types of complexity in megaprojects: structural and dynamic. Their study of two megaprojects revealed that successful management of this mix of complexity and uncertainty involved adopting structures and processes that foster collaborative behaviours. It also highlighted the importance of being adaptive and responsive, as well as the use of innovative approaches and digital technologies.

Box 2.12 Case study – the uncertainty management house.

In their investigation of six exploratory projects in the automotive and chemical industries, Gomes et al. (2019) uncovered significant findings in identifying three novel categories of uncertainty faced by project managers: primitive, structural, and elementary.

Building on this classification, Gomes et al. (2019) proposed a structured, multi-level approach for managing these uncertainties, which they termed the "uncertainty management house". This framework offers a more systematic method for dealing with uncertainties in exploratory projects, aiming to enhance management fit and alignment. This approach is particularly relevant for program management or the management of a portfolio of projects.

A notable aspect of this study is its focus on understanding actual practices and strategies employed in transitioning from uncertainty to risk. This perspective marks a departure from previous research, which primarily concentrated on developing rational models and tools for predicting and controlling behaviour. By emphasising practical, real-world applications and strategies, the study contributes valuable insights into effective project management in the context of handling uncertainty in complex, exploratory projects.

Box 2.13 The concept of ambidextrous organisations to navigate chaos and order.

In her study of multi-project firms, Geraldi (2008) aligned with the concept of ambidextrous organisations, positing that successful organisations are those that can navigate the co-existence of chaos and order. Geraldi's work led to the development of a framework for mapping four organisational archetypes. This framework is based on the tendency towards "chaotification" of order and bureaucratisation of chaos. It provides a comprehensive view of a project management firm's organisational design, highlighting the spectrum of complexity and flexibility inherent in these organisations.

This framework is instrumental in delineating the organisational architecture of megaprojects. The framework serves as a tool for identifying both permanent and temporary organisations and how they adjust and reconfigure to support different phases of a megaproject. It is especially beneficial for strategising the delivery of megaprojects and clarifying the roles and relationships within multi-organisational layers.

Power and politics

The role of power dynamics in the success of megaprojects has been increasingly recognised (Soderlund et al., 2017; van Marrewijk, 2018; Clegg et al., 2017; Clegg, 2019). Megaprojects are not just technical endeavours, they are complex social constructs shaped significantly by power relations. These projects are managed and delivered within a framework of power dynamics, deeply influenced by social construction (Clegg & Kreiner, 2013).

Given this context, van Marrewijk (2018) argues for the necessity of addressing power relations in the analysis of megaprojects. This includes a thorough consideration of the social, cultural, and political contexts and processes that influence megaproject management.

While traditional mechanistic approaches still prevail in much of the current project management literature, there's a growing shift in focus. More attention is now being paid to the aspects of social interaction, project culture, and power relations. This shift is evident in the works of scholars like Cicmil and Hodgson (2006), van Marrewijk (2017), and Ninan et al. (2019).

In his early case study on power dynamics within projects, Clegg (1975) revealed that decision-making processes were often fraught with tension and conflict. He observed that these processes frequently involved acrimonious disputes and "language games", which unfolded in various settings, including both formal and informal project meetings.

In most construction projects, the contract and its related documents are traditionally viewed as the cornerstone, clearly defining the roles and responsibilities of the project team members. These documents are intended to serve as a guiding framework for project execution. However, the reality is often more complex.

Skills professionals in the industry frequently encounter large and intricate bodies of contract documentation. These documents, while comprehensive, can contain ambiguities that professionals must interpret and navigate. This situation often leads to the contract documentation becoming a battleground where ambiguities are translated into contested actions. Such scenarios highlight the significance of interpretation and negotiation skills in managing and resolving disputes, and in effectively guiding project progress amidst complex contractual frameworks.

Box 2.14 Case study – tactical clustering in megaproject management.

Drawing on theories of power, particularly an organisational power-based approach, Ninan et al. (2019) provided insightful observations from a case study of an Indian megaproject. This research uncovered how the management of the project unfolded through a variety of strategies, which materialised through a process they termed "tactical clustering".

Focusing on the role and influence of external stakeholders in megaprojects, Ninan et al. (2019, p. 626) identified a critical aspect of project management. They found that in situations where formal contracts and standards were not sufficient to govern these stakeholders, project teams often had to "muddle through with short-term tactics". These tactics were not random but were instead enabled by the availability of resources embedded with existing power structures.

One of the key contributions of this study lies in its illumination of how various stakeholders, not formally integrated into the decision-making process, can still significantly influence the outcomes of megaprojects. These stakeholders, driven by their vested interests, can shape project decisions and trajectories, sometimes subtly, and at other times more overtly. Ninan et al. (2019) effectively demonstrate the nuanced and often informal ways in which stakeholder power dynamics play out in the context of megaprojects, impacting their success and direction.

Megaprojects often present a unique landscape where both project team members and external stakeholders can pursue individualistic interests through the exercise of power (Lehrer & Laidley, 2008). This dynamic can sometimes lead to the prioritisation of personal interests over the collective goals of the project.

Gil (2015) has shown that the subpar performance of megaprojects can often be traced back to the need to accommodate the demands of various stakeholders. This compliance can sometimes divert the project from its intended objectives, leading to inefficiencies or misalignments with the original project goals. Keane (2012, p. 660) further elaborates on this dynamic by describing the power relations in megaprojects as being shrouded in a "canopy of multiple realities". These include hidden agendas and are often shielded by efforts to maintain silence, relying on a façade of public rhetoric and outward appearances.

While the use of overt forms of power is more visible and easily identifiable in project settings, covert power operates differently. It tends to be more elusive, often hidden, and protected within the realms of taken-for-granted practices. The cover exercise of power can significantly influence project outcomes, yet its identification and understanding are challenging due to its concealed nature.

Given the substantial impact that covert power can have on the performance and direction of megaprojects, it's crucial to understand the various ways it is exercised. This understanding can help project managers and stakeholders navigate the complex power dynamics and ensure that project goals are achieved effectively, keeping in mind the broader interests of the project rather than individual or subgroup interests.

Management and organisation

The issue of cost overruns and underwhelming operational outcomes in megaprojects has been a focal point of scholarly interest for over four decades (Siemiatycki, 2016; Flyvbjerg et al., 2003; Altshuler & Luberoff, 2003). Historically, research in the management and organisation disciplines regarding megaprojects has largely concentrated on what is often termed "pathologies", which primarily involve identifying the causes of cost overruns and the overall poor performance of these large-scale projects.

One particularly influential work in this area is Flyvbjerg, Holm, and Buhl's seminal article titled "Cost underestimation in public projects: Error or lie?" published in the *Journal of the American Planning Association* in 2002. This article has been regarded as a classic in the field of megaproject management and has significantly shaped the discourse on this subject. Since its publication, the article has been cited numerous times, evidencing its profound impact on the field. It has spurred an intense and ongoing research agenda focused on understanding and addressing persistent issues of cost overruns and failures in project delivery.

Box 2.15 Case study – deception and political-economic interests contributing to cost overruns in megaprojects.

The research trajectory focusing on systemic cost overruns in megaprojects was further expanded by a significant study conducted by Flyvbjerg (2009). This study examined cost data from 258 transportation infrastructure projects, providing compelling evidence of widespread cost overruns. Their research, which remains a seminal work in the field of megaproject studies, found an overwhelming majority – nine out of ten projects – experienced cost overruns. This was observed across projects in 20 nations spanning five continents, emphasising the global and pervasive nature of this issue.

A key aspect of this research agenda was its departure from traditional technical explanations commonly put forth by project management researchers, Instead, Flyvbjerg (2009) delved into alternative perspectives, focusing on the roles of deception and political-economic interests in contributing to cost overruns in megaprojects. This approach marked a shift towards understanding the deeper, often non-technical factors that influence project costs and performance.

Political-economic interests

The research within the management and organisation domain, particularly focusing on the pathologies of megaprojects, consistently points out that a primary challenge in such projects is the significant inaccuracies in forecasts related to costs, demands, and other impacts. Flyvbjerg, a leading figure in megaproject research, delved into the reasons behind these inaccuracies and categorised the explanations found in the literature into three broad types: technical complexity, psychological issues, and political-economic decisions.

Technical complexity has been identified as the most common explanation for forecasting inaccuracies in megaprojects (Morris & Hough, 1987; Wachs, 1989; Flyvbjerg et al., 2002, Flyvbjerg, 2005). According to this viewpoint, inaccuracies typically arise from honest mistakes, insufficient data, the incompetency of forecasters, and flawed forecasting techniques. This explanation suggests that the path to improved accuracy lies in the development of more advanced forecasting models, enhanced data collection, and the employment of more skilled and experienced forecasters.

Flyvbjerg (2009) critically evaluated and ultimately challenged the technical explanations commonly given for inaccuracies in forecasting, particularly in the context of megaprojects. His rejection of these explanations is grounded in two key observations:

1 *Distribution errors in forecasts*: Flyvbjerg pointed out that if technical complexity were the primary cause of forecasting inaccuracies, then the distribution of errors in these forecasts should be less biased. However, the actual data on

forecasting inaccuracies show a statistical distribution that does not align with what one would expect from mere technical errors. This discrepancy suggests that factors other than technical difficulties might be playing a significant role in the inaccuracies observed.

2 *Lack of improvement over time*: The second reason for rejecting technical explanations relates to the expectation of improved accuracy over time. If inaccuracies were primarily due to technical issues, then over time, as errors and their sources are identified and addressed – through better data collection, refined forecasting methods, etc. – the accuracy of forecasts should improve. However, Flyvbjerg observed that despite advancements in data sources and forecasting methods, there has been no noticeable improvement in the accuracy of forecasts. This lack of progress indicates that the root causes of inaccuracies might lie beyond technical aspects.

Box 2.16 The concepts of planning fallacy and optimism bias in megaproject decision-making.

The poor performance of megaprojects in terms of forecasting inaccuracies has been explored from a psychological perspective (Kahneman & Lovallo, 1993; Lovallo & Kahneman, 2003). They introduce concepts like the planning fallacy and optimism bias to explain these inaccuracies.

1 *Planning fallacy*: Kahneman and Lovallo (1993) described the planning fallacy as a tendency for forecasters to be overly optimistic when making predictions. This optimism is not grounded in rational calculations of benefits, disadvantages, and probabilities. Instead, it is based on what they term "delusional optimism". This mindset leads forecasters to overestimate the benefits of a project while underestimating the costs and challenges.
2 *Optimism bias*: Lovallo and Kahneman (2003) further elaborate on optimism bias, which is closely related to the planning fallacy. This bias involves a kind of wishful thinking, where forecasters imagine the most favourable scenarios for project success, often disregarding or minimising the potential for errors, miscalculations, and unforeseen obstacles.

In such cases, the forecasters' decision-making process is skewed by an overly positive view of the future, leading to unrealistic forecasts.

Flyvbjerg (2009) delved into the psychological underpinnings of over-optimistic behaviour in megaproject forecasting, attributing it to cognitive biases – errors in the way the mind processes information. Despite the prevalence of these biases, there has been an assumption that they can be mitigated or even eliminated through simple reality checks. This would theoretically reduce the likelihood of project stakeholders making hasty and unprofitable investments due to increased accuracy in forecasting. However, empirical evidence, especially from studies like Flyvbjerg (2005), suggests that this assumption hasn't held true in practice.

There is substantial evidence supporting the existence of optimism bias. However, a notable issue is that much of this data comes from simpler, non-professional settings. Considering the increasing level of professional expertise, scientific analysis, critical assessment, and peer review involved in forecasting for megaprojects, it seems implausible that a whole profession of forecasting experts would persistently repeat the same mistakes over decades without learning from past experiences.

Flyvbjerg (2005) argue that for forecasters to maintain an optimism bias over the 70-year period their study covers, without learning from past errors, would suggest a degree of unprofessionalism and perpetual optimism. This view challenges the notion that optimism bias alone is the primary cause of inaccuracies in megaproject forecasts.

Box 2.17 Strategic misrepresentation for furthering agendas.

In his 2009 work, Flyvbjerg (2009) shifted the focus from usual technical and psychological reasons to a new perspective for understanding cost overruns in large-scale projects. He suggested that those in charge of planning such projects might intentionally overstate benefits and understate costs in their forecasts. This approach detailed further in Flyvbjerg's studies from 2002, 2005, and 2009, posits that project promoters often present overly optimistic scenarios, downplaying potential failures. According to Flyvbjerg (2009) this "strategic misrepresentation" is often driven by political and organisational pressures within the sponsoring entities. It serves as a tactic for various groups or departments to further their own agendas. For instance, in projects sponsored by government agencies, there might be competition for limited funding or a struggle for influence, prompting promoters to manipulate information to achieve broader objectives or goals.

Strategic misrepresentation is often viewed by those in political and economic sectors as a pragmatic approach to achieving their goals. In this context, accuracy in forecasting and business case development takes a back seat to the primary objective of getting projects approved and funded. This perspective aligns with the viewpoint of Wachs in 1989, who observed that the most effective planners are often those who can disguise their advocacy as scientific or technical objectivity. This political-economic interpretation explains why many projects fail: agencies or departments too readily endorse what they claim are superior solutions early in a project's lifecycle, without a fair and thorough evaluation of alternatives. Consequently, the strategic misrepresentation practised by project promoters can overshadow more rational methods, such as the systems analysis and system dynamics approaches discussed earlier.

Political power and deception

The importance of accurate information in the decision-making process for megaprojects is well recognised. However, existing research, including works by Flyvbjerg et al. (2002; Flyvbjerg, 2005, 2009) and Ronnle (2017), has revealed that the analyses and reports prepared for these decisions often serve purposes beyond merely informing. They are frequently used to case projects in a more favourable light or to symbolically justify decisions already made. This phenomenon was noted even before Flyvbjerg et al.'s landmark 2002 study on cost underestimation, indicating a longstanding interplay between information use and political-economic interests in the megaproject decision-making environment.

Box 2.18 Case study – the opportunistic use of political power.

In his 1998 study of the Aalborg project, a complex urban development in Denmark, Flyvbjerg illustrated how political power can be opportunistically used to achieve individual objectives. While bureaucracies are often seen as resistant to change, the Aalborg project serves as an example that significant shifts can occur in large organisational structures when there's a change in the power dynamics. This project, aimed at urban renewal, land use, traffic, and environmental issues, unfortunately led to "environmental degradation and social distortion", as Flyvbjerg noted (1998, p. 318). His case study provides a clear picture of how certain stakeholders wielded political power, the nature of the relationships involved, and the consequent effects on the project's outcomes.

The case study also reveals the existence of an informal, clandestine "council" comprising business and government representatives. In this council, decisions concerning the Aalborg project, as well as other policies and plans relevant to the business sector, were collaboratively negotiated, and implemented in a corporative manner, prior to any external input. The prominence of business interests in the Aalborg project was further reinforced by the concerted backing of local media and law enforcement. Democratically elected government bodies found their influence markedly diminished in this context. Similarly, other community groups, apart from the business community, and the general public, had limited impact on the final outcomes of these decisions.

Ronnle (2017) supports Flyvbjerg's earlier findings, indicating that in decision-making for megaprojects, the use of information is heavily influenced by political factors, and opportunistic behaviour is common. It's notable that much of the existing research on power dynamics and deception focuses primarily on high-level decisions made by politicians and planners, usually at the initial stages of projects. However, while deceptive practices of contractors have been associated with poor performance in megaprojects, there is limited understanding of the specific ways these practices are carried out.

Improving accountability

In response to the recognition of political-economic motives and "rent-seeking" behaviours contributing to poor performance in megaprojects, various mechanisms have been suggested to enhance accountability. These include implementing clear policies, procedures, and institutional structures. There has been a considerable focus on creating measures and mechanisms to ensure that planning analysis is conducted transparently and with methodological integrity.

Bruzelius et al. (2002) suggested four specific measures to increase accountability which include:

- Transparency
- Performance specifications
- Explication of regulatory regimes
- Involvement of risk capital

Flyvbjerg et al. (2003) laid out a detailed set of recommendations to improve the management of megaprojects, centred around three core principles. First, they proposed a legal mandate for comprehensive risk analysis as a part of the initial feasibility studies for all large-scale projects, Second, the roles of public sector actors, especially politicians, should be clearly demarcated, focusing primarily on setting, and auditing public interest objectives. The third principle emphasises the need for increased participation from various stakeholder groups and civil society to enhance transparency and accountability.

However, while these measures for improving transparency and accountability are theoretically sound, they may not always align with the realities of policy and planning. Political processes are often governed by irrationalities, and deviations from established rules are common. Merely advocating for well-defined policies and scientific approaches might not significantly alter the attitudes and behaviours of those involved in the political decision-making processes of megaprojects.

Urban and environmental planning

The increasing significance of megaprojects in urban and environmental planning is evident from the dedicated attention they have received in academic literature, such as the 2017 special issue on "Critical Urban Infrastructure" in the journal "*Urban Policy and Research*". Traditionally viewed as the realm of engineering and high-level politics, Steele and Legacy (2017) pointed out that urban infrastructure is a relatively unexplored area in critical urban research, policy, and practice.

Despite this, there has been a convergence of research across various disciplines within urban and environmental planning. This interdisciplinary approach incorporates themes commonly found in management and organisation and project management studies. Key areas of focus include participatory decision-making, effective stakeholder management, governance, and addressing political challenges arising from the increased emphasis on infrastructure development.

Participatory decision-making

The field of environmental planning has seen notable advancements, particularly in the development of participatory decision-making processes aimed at empowering users and involving diverse stakeholder groups and civil society. This shift towards more inclusive planning practices recognises the importance of various perspectives in shaping environmental policies and projects.

However, this move towards greater participation has also uncovered challenges, particularly in ensuring adequate representation of different users in the decision-making process. Different interests between stakeholder groups and the associated complexity in achieving consensus have been a common theme within the environmental planning literature (Buchy & Race, 2001; Innes & Booher, 2004).

Furthermore, the environmental planning literature consistently points to the varying beliefs and values of individuals regarding environmental issues (Michelson, 1980; Rapoport, 1994; Wilson, 1996). These differences can lead to conflicts and pose significant challenges in collaborative planning efforts.

Addressing the challenges of representation and conflict in participatory environmental planning, significant research has focused on conflict resolution techniques (Moore, 1986) and consensus-building and participation methods (Habermas, 1981; Sanoff, 2000; Innes & Booher, 2004) based on effective communication. Most of the participatory models developed since the 1980s draw heavily on Habermas' concept of communicative rationality from his 1981 theory of communicative action. Habermas argues that true collaboration arises from a form of rationality where meanings, values, understandings, and knowledge are collectively created through deliberative dialogue. He envisions an ideal scenario of domination-free communication among stakeholders, where agreements are reached not through coercion but through a shared conviction and a sense of equality among all parties involved.

Stakeholder management

Stakeholder management is an increasingly important strategy in project development, facilitating active interaction among stakeholders throughout the project lifecycle. This approach is aimed at fostering collective discourse about project needs and expectations. Gaining traction in the late 1990s within the construction research community, stakeholder management aligns with strategic organisational planning trends and has recently resurfaced as a vital tool for achieving social, environmental, and economic goals in projects.

Various tools and methods have been developed to manage stakeholders' expectations and requirements across different types of projects. A typical management approach, as described by Oyegoke (2010), includes six key principles: early identification of stakeholders, identification of potential conflict areas, education of stakeholders on potential risks and harms and how they are mitigated and potential gains, engagement with stakeholders, involvement of other entities particularly government to demonstrate that due process is followed, and management of the process.

Box 2.19 Stakeholder analysis – the power–interest matrix.

Stakeholder analysis is a crucial component in stakeholder management. It involves identifying and characterising stakeholders based on their attributes, interrelationships, and interests concerning a specific issue (Manowong & Ogunlana, 2010). A fundamental tool used in stakeholder analysis is the power–interest matrix, which helps in identifying stakeholders based on their level of power and interest in a project.

The power of a stakeholder is defined as their capacity to influence project outcomes, while their interest is their level of desire to influence these outcomes. Using this matrix, stakeholders can be classified as either key or non-key and can be categorised into four groups (Leung & Olomolaiye, 2010):

- Group 1, identified as key player stakeholders, is characterised by high levels of power and interest in the project. This group is crucial due to their significant ability to influence decision-making and project outcomes. Stakeholders in this category typically include clients, their representatives, consultants, contractors, suppliers, and any parties delegated with authority by the client to execute project activities.
- Group 2, known as "keep satisfied" stakeholders, is characterised by a high level of power but a low level of interest in the project. This group may not be directly involved in the project's day-to-day activities, which accounts for their lower interest. However, their ability to exert significant power over the project cannot be overlooked. An example of stakeholders in this category includes local regulatory bodies responsible for approving project documentation. While they might not be actively engaged in the project's operational aspects, their decisions and actions, such as reviewed regulations or additional requirements, can have substantial impacts.

- Group 3, labelled as "keep informed" stakeholders, is characterised by a high level of interest in the project but possess relatively low power over decision-making. These stakeholders often have a significant stake in the project's outcomes, which drives their keen interest. Despite their limited direct influence, stakeholders in this group can affect public perception of the reputation of the client or project. Their opinions and reactions can influence broader community sentiments or consumer attitudes, which can be crucial for the project's overall success and acceptance.
- Group 4, termed as "minimal effort" stakeholders, is characterised by both low levels of interest and power in relation to the project. These stakeholders pose the least risk in terms of impact on the project compared to other groups, owing to their limited engagement and influence. While they typically require the least attention and management effort, certain situations may elevate their relevance. In such cases, specific coping strategies might become necessary to manage their newfound concerns or influence.

Effective stakeholder management hinges on the ability to accurately assess and understand the varying levels of power, influence, agendas, and potential implications posed by different stakeholders. These aspects can be either overt or covert. Power dynamics within a project often prompt stakeholders to adopt various strategies or tactics to position themselves advantageously, influencing project outcomes.

Central to the contributions in collaborative planning and stakeholder management is the focus on modes of interaction between stakeholders. This includes how problems within the project are analysed, synthesised, and evaluated, and the processes governing these activities. The way stakeholders interact, communicate, and engage with each other plays a crucial role in determining the project's success. The underlying assumption of these approaches is that employing specific techniques and methods will aid in navigating the diversity of interests and the conflicts that inherently arise in projects.

The infrastructure turn

The last decade has witnessed what has been described as an "infrastructure turn" (Dodson, 2017), which has serious implications for urban scholarship. This shift, first identified in the Australian urban policy context during the 2000s, has since gained momentum globally. There has been a concerted effort worldwide to stimulate infrastructure development through various international frameworks. As a result, megaprojects, which were relatively rare two decades ago, have now become a common feature in the landscape of urban development. Set against this backdrop is an expanding body of research that delves into the myriad challenges associated with the planning and delivery of these large-scale projects.

Over the past 15 years, there has been a rise in the development of consensus-based processes to support inclusive public dialogue, mirroring the participatory models discussed earlier. In Australia, there has been a significant shift from a traditionally instrumental rationality, which prioritised professional expertise, to a more inclusive approach that values the input of civil society. Inclusive deliberation has become a key feature of the current inclusive governance environment (Legacy et al., 2017).

Despite these advancements, concerns persist about the influence of politics and power dynamics in contemporary decision-making processes (Purcell, 2009; Žižek, 1999). These concerns reflect an ongoing debate about the extent to which governance can genuinely mitigate power imbalances and political influences in decision-making. The focus of much of this recent research has been predominantly on urban infrastructure, particularly transport and road network projects, rather than on construction megaprojects.

The emergence of smart city strategies, closely intertwined with digital infrastructure and public sector information (PSI), represents another significant area of development. The OECD estimated the value of PSI data at EUR 32 billion for the European market alone (Ubaldi, 2013). This value stems from innovative organisational and management approaches, encompassing diverse areas such as weather forecasts, traffic management, crime statistics, and improved transparency of government functions (Ubaldi, 2013).

The global rise of smart city initiatives, however, is not without its challenges. One critical aspect that requires attention is the governance implications of digital infrastructure in the era of smart cities (Barns et al., 2017). While smart cities differ significantly from conventional cities in terms of planning and delivery, they share many challenges related to political-economic interests and the dynamics of power in decision-making.

Current challenges

The discourse on megaprojects, through an examination of various trends, models, and events in this chapter, reveals key shortcomings in current approaches to enhancing megaproject performance. These shortcomings highlight areas for further understanding:

1 *Lack of empirical narratives*: There is a noticeable absence of detailed empirical accounts that capture the dynamic, spontaneous governing occurring in megaprojects as they unfold.
2 *Human aspect of governance overlooked*: Most project governance frameworks focus predominantly on formal roles and regulations. There is a limited understanding of the human side of governance, particularly the diverse forms of power, authority, and subjectivity that play a significant role in the execution of megaprojects.
3 *Navigating institutional complexity*: A critical gap identified is the insufficient understanding of the dynamics, institutional plurality, and complexity navigated

by various project team members in their daily decision-making and power plays within megaprojects.

This book addresses these shortcomings by presenting an innovative framework that transcends the conventional focus on formally prescribed mechanisms in megaproject management to also consider the informal power relations that significantly influence decision-making. Moving beyond the traditional emphasis on rational forms of participation and stakeholder management, it focuses on a more pragmatic and realistic approach. This approach recognises the critical role of power in the decision-making processes of megaprojects.

Drawing inspiration from the work of Soderlund et al. (2017), the book emphasises the importance of understanding megaprojects not just as final products or outcomes, but as dynamic processes of organisation. It focuses on the stories of individuals involved in these projects, viewing them as evolving organisational entities and action localities where politics and power intermingle.

Aligned with the projects-as-practice perspective (Cicmil et al., 2006), this book contributes to a growing body of work (e.g. van Marrewijk, 2015). It aims to demystify the "black box" of megaproject management by examining the micro-processes through which clients and project team members enact and translate governance frameworks into practical reality.

Specifically, this book addresses critical questions that have not been adequately explored in existing literature. These questions delve into how team members and clients in megaprojects incorporate social norms, cultural beliefs, and local expectations into the structure and processes of megaprojects. It seeks to shed light on how norms and work practices are established and evolve within these projects, how managers navigate institutional contradictions and conflicts, and the characteristics of institutional entrepreneurs within megaprojects, including the strategies they employ and their effectiveness. By addressing these questions, this book provides a deeper understanding of the complex social, cultural, and political dynamics at play in megaprojects and offers practical insights for improving their management and outcomes.

Chapter summary

This chapter has synthesised the evolution and current state of megaproject research since the 1990s, a period marked by a global pursuit of enhanced megaproject performance by academics, industry practitioners, and policymakers. Key gaps in the literature were identified to highlight its implications for the approach undertaken in this book. This book aims to enrich the existing discourse on megaprojects in three pivotal ways: first, by focusing on actual governing processes, power relations and deviations encountered by clients and project team members. Second, by examining the multi-faceted nature of power, illuminating how it is created, distributed, and exercised within the context of megaprojects. Third, by examining the taken-for-granted everyday decision-making practices occurring at the micro level, as opposed to the high-level macro decisions of ideal typification.

References

Abdul-Aziz, A., Wong, S., & Awil, A. (2008). Formulating a holistic model to describe international diversification of construction companies. *Building abroad: Procurement of construction and reconstruction projects in the international context*, Groupe de recherché if – grif, Université de Montréal, http://www.grif.umontreal.ca/pages/conferencegrif08.htm (Accessed: April 2009).

Altshuler, A., & Luberoff, D. (2003). *Megaprojects: The changing politics of urban public investment*. Washington, DC: Brookings Institution.

Antoniadiis, D.N., Edum-Fotwe, F.T., & Thorpe, A. (2011). Socio-organo complexity and project performance. *International Journal of Project Management*, 29(7), 808–816.

Association for Project Management (APM). (2002). Directing change; A guide to governance of project management, http://www.apm. org.uk.

Atkinson, R., Crawford, L., & Ward, S. (2006). Fundamental uncertainties in projects and the scope of project management. *International Journal of Project Management*, 24, 687–698.

Baccarini, D. (1996). The concept of project complexity: A review. *International Journal of Project Management*, 14(4), 201–204.

Barns, S, Cosgrave, E., Acuto, M., & Mcneill, D. (2017). Digital infrastructures and urban governance. *Urban Policy and Research*, 35(1), 20–31.

Biesenthal, C., Clegg, S.D., Mahalingam, A., & Sankaran, S. (2018). Applying institutional theories to managing megaprojects. *International Journal of Project Management*, 36(1), 43–54.

Boateng, P., Chen, Z., Ogunlana, S., & Ikediashi, D. (2012). A system dynamics approach to risks description in megaprojects development. *Organisation Technology Management Journal*, 3(3), 593–603.

Bon, R., & Crosthwaite, D. (2000). *The future of international construction*. London: Thomas Telford.

Brady, T., & Davies, A. (2014). Managing structural and dynamic complexity: A tale of two projects. *Project Management Journal*, 45(4), 21–38. DOI: 10.1002/pmj.21434

Bridge, A., In, S., & Rowlinson, S. (2015). Models for engaging public-private partnerships in civil infrastructure projects: A case of 'having your cake and eating it too'? In G. Macht, & C. Dossick (Eds.), *Proceedings of the 12th Engineering Project Organization Conference*. Engineering Project Organization Society (EPOS), pp. 1–14.

Briscoe, G., Dainty, A., Millett, S., & Neale, R. (2004). Client-led strategies for construction supply chain improvement. *Construction Management and Economics*, 22(2), 193–201.

Brockmann, C., & Girmscheid, G. (2007). Complexity of megaprojects, CIB World Building Congress: Construction for development: 14–17 May 2007, Cape Town International Convention Centre, South Africa.

Brookes, N., Chen, Z., & Storto, C. (2014). Editorial special issue of Infrastructure megaprojects. *International Journal of Architecture, Engineering and Construction*, 3(2), 1–2.

Bruzelius, N., Flyvbjerg, B., & Rothengatter, W. (2002). Big decisions, big risks. Improving accountability in mega projects. *Transport Policy*, 9(2), 143–154.

Buchy, M., & Race, D. (2001). The twists and turns of community participation in natural resource management in Australia: What is missing? *Journal of Environmental Planning and Management*, 44(3), 239–308.

Busby, J., & Hughes, E. (2004). Projects, pathogens and incubation periods. *International Journal of Project Management*, 22(5), 425–434.

Chan, P., & Cooper, R. (2007). What makes a leader in construction? An analysis of leaders in the UK construction industry, political science, 14–17 May 2007, Cape Town International Convention Centre, South Africa.

Chan, E., & Tse, R. (2003). Cultural considerations in international construction contracts. *Journal of Construction Engineering and Management*, 129(4), 375–381.

Chi, C., & Javernick-Will, A. (2011). Institutional effects on project arrangement: High-speed rail projects in China and Taiwan. *Construction Management and Economics*, 29(6), 595–611.

Christensen, T. (2007). The Norwegian front-end governance regime of major public investment projects—A theoretically based analysis and evaluation (Preliminary version of Concept Research Programme report).

Cicmil, S., & Hodgson, D. (2006). New possibilities for project management theory: A critical engagement. *Project Management Journal*, 37(3), 111–122.

Cicmil, S., Williams, T., Thomas, J., & Hodgson, D. (2006). Rethinking project management: Researching the actuality of projects. *International Journal of Project Management*, 24(8), 675–686.

Clarke, T. (2004). Theories of governance—Reconceptualizing corporate governance theory after the Enron experience. In T. Clarke (Ed.), *Theories of corporate governance: The philosophical foundations of corporate governance* (pp. 1–30). Abingdon: Routledge.

Clegg, S.R. (1975). *Power, rule and domination.* London: Routledge and Kegan Paul.

Clegg, S. (2019). Governmentality. *Project Management Journal*, 50(3), 266–270.

Clegg, S.R., & Kreiner, K. (2013). Power and politics in construction projects. In N. Drouin, R. Müller, & S. Sankaran (Eds.), *Novel approaches to organizational project management research: Translational and transformational* (pp. 268–293). Copenhagen: CBS Press.

Clegg, S., Sankaran, S., Biesenthal, C., & Pollack, J. (2017). Power and sensemaking in megaprojects. In B. Flyvbjerg (Ed.), *The Oxford handbook of megaproject management* (pp. 238–258). Oxford: Oxford University Press.

Crawford, L., Cooke-Davies, T., Hobbs, B., Labuschagne, L., Remington, K., & Chen, P. (2008). Governance and support in the sponsoring of projects and programs. *Project Management Journal*, 39, 43–55.

Crawford, L., & Cooke-Davies, T. (2009). Project governance: The role and capabilities of the executive sponsor. *Project Perspectives*, 31, 66–74.

Desjardins, J. (2017). Nine of the world's largest megaprojects that are under construction. *Business Insider*, January 11, 2017.

Dewulf, G., Garvin, M.J., & Duffield, C. (2016). Multinational comparison of the tension between standards and context in PPP. In A. Roumboutsos (Ed.), *Public private partnerships in transport: Trends and theory* (pp. 267–291). New York: Routledge Studies in Transport Analysis.

Dewulf, G., & Garvin, M. (2020). Responsive governance in PPP projects to manage uncertainty. *Construction Management and Economics*, 38(4), 383–397.

Ding, Y., & Poon, S. (2009). Mega-challenges: Programming management for event projects. In *International symposium*, CIBW096 Future trends in architectural management, Nov 2–4, Taiwan.

Dodson, J. (2017). The global infrastructure turn and urban practice. *Urban Policy and Research*, 35(1), 87–92.

Dunning, J. (2000). The eclectic paradigm as an envelope for economic and business theories of MNE activity. *International Business Review*, 9(2), 163–190.

Ebrahimnejad, A., & Tavana, M. (2014). A novel method for solving linear programming problems with symmetric trapezoidal fuzzy numbers. *Applied Mathematical Modelling*, 38(17–18), 4388–4395.

Eisenhardt, K. (1989). Building theories from case study research. *Academy of Management*, 14(4), 532–550.
Fiori, C., & Kovaka, M. (2005). Defining megaprojects: Learning from construction at the edge of experience. ASCE Construction Research Congress: Broadening perspectives, April 5–7, California.
Flyvbjerg, B. (1998). *Rationality and power: Democracy in practice*. Chicago, IL: The University of Chicago Press.
Flyvbjerg, B. (2005). Design by deception: The politics of megaproject approval. *Harvard Design Magazine*, 22, 50–59.
Flyvbjerg, B. (2009). Survival of the unfittest: why the worst infrastructure gets built—and what we can do about it. *Oxford Review of Economic Policy*, 25(3), Autumn 2009, 344–367.
Flyvbjerg, B. (August 1, 2006). From Nobel prize to project management: Getting risks right. *Project Management Journal*, 37(3), 5–15.
Flyvbjerg, B. (2014). What you should know about megaprojects and why: An overview. *Project Management Journal*, 45(2), 6–19.
Flyvbjerg, B. (2017). *The Oxford handbook of megaproject management*. Oxford: Oxford University Press.
Flyvbjerg, B., & Turner, R. (2017). Do classics exist in megaproject management? Introduction to special issue on classics in megaproject management. *International Journal of Project Management*, 36, 334–341.
Flyvbjerg, B., Holm, M.K.S., & Buhl, S.L. (2002). Underestimating costs in public works project: Error or lie? *Journal of American Planning Association*, 68(3), 279–295.
Flyvbjerg, B., Bruzelius, N., & Rothengatter, W. (2003). *Megaprojects and risk: An anatomy of ambition*. Cambridge: Cambridge University Press.
Garland, D. (1997). Governmentality and the problem of crime: Foucault, criminology and sociology. *Theoretical Criminology*, 1(2), 173–214.
Geraldi, J.G. (2008). The balance between order and chaos in multi-project firms: A conceptual model. *International Journal of Project Management*, 26(4), 348–356.
Giezen, M., Salet, W., & Bertolini, L. (2015). Adding value to the decision-making process of mega projects: Fostering strategic ambiguity, redundancy, and resilience. *Transport Policy*, 44, 169–178.
Gil, N. (2009). Developing cooperative project-client relationships: How much to expect from relational contracts. *California Management Review*, 51(2), 144–169. DOI: 10.2307/41166484
Gil, N.A. (2015). Sustaining highly-fragile collaborations: A study of planning mega infrastructure projects in the UK. Working Paper.
Gomes, L., Brasil, V., & Paula, R. (2019). Proposing a multilevel approach for the management of uncertainties in exploratory projects. *Project Management Journal*, 50(5), 554–570.
Grilo, L., Melhado, S. Silva, R., Edwards, P., & Hardcastle, C. (2007). International building design management and project performance: Case study in Sao Paulo, Brazil. *Architectural Engineering and Design Management*, 3, 5–16.
Habermas, J. (1981). *The theory of communicative action: Vol 1: Reason and the rationalization of society*. London: Polity Press.
Hagan, G., Baltov, M., & Brookes, N. (2012). Megaproject management: The state of the art. *Organisational Technology Management Construction*, 4, 567–569.
Hirschman, A.O. (1995). *Development projects observed* (2nd ed.). Washington, DC: Brookings Institution.
Hobday, M. (1998). Product complexity, innovation and industrial organisation. *Research Policy*, 26(6), 689–710.

Hu, Y., Chan, A.P., Le, Y., & Jin, R.Z. (2015). From construction megaproject management to complex project management: Bibliographic analysis. *Journal of Management in Engineering*, 31(4), 1–11.
Innes, J., & Booher, D. (2004). Reframing public participation: Strategies for the 21st century. *Planning Theory and Practice*, 5(4), 419–436.
Jacoby, S. (2005). Corporate governance and society. *Challenge*, 48(4), 69–87.
Javernick-Will, A., & Scott, W.R. (2011). Who needs to know what? Institutional knowledge and global projects. In W.R. Scott, R.E. Levitt, & R.J. Orr (Eds.), *Global projects: Institutional and political challenges* (pp. 247–276). Cambridge: Cambridge University Press.
Javernick-Will, A., & Levitt, R. (2009). Acquiring local knowledge for international projects. Construction Research Congress 2009 Seattle, April 5–7.
Jones, C., Heserly, W.S., & Borgatti, S.P. (1997). A general theory of network governance: Exchange conditions and social mechanisms. *Academy of Management Review*, 22, 911–945.
Kahneman, D., & Lovallo, D. (1993). Timid choices and bold forecasts: A cognitive perspective on risk taking. *Management Science*, 39(1), 17–31.
Keane, J. (2012). Silence and catastrophe: new reasons why politics matters in the early years of the twenty-first century. *The Political Quarterly*, 83(4), 660–668.
Khallaf, R., Naderpajouh, N., & Hastak, M. (2018). A systematic approach to develop risk registry frameworks for complex projects. *Built Environment Project and Asset Management*, 8(4), 334–347.
Klakegg, O.J., Williams, T., Magnussen, O.M., & Glasspool, H. (2008). Governance frameworks for public project development and estimation. *Project Management Journal*, 39, S27–S42.
Klakegg, J., Williams, T., & Shiferaw, A. (2016). Taming the 'trolls': Major public projects in the making. *International Journal of Project Management*, 34(2), 282–296.
Kobayashi, K., Rashid, K., Ofori, G., & Ogunlana, S. (2009). *Joint ventures in construction.* London: Thomas Telford.
Kumaraswamy, M., & Morris, D. (1998). Build-operate-transfer-type procurement in Asian megaprojects. *Journal of Construction Engineering and Management*, 128(2), 93–102.
Kusuma, I. (2014). The cultural ecosystem of megaprojects: The interconnectedness of organizational elements and their wider institutional contexts. *International Journal of Architecture, Engineering and Construction*, 3(2), June 2014, 82–97.
Legacy, C., Curtis, C., & Scheurer, J. (2017). Planning transport infrastructure: Examining the politics of transport planning in Melbourne, Sydney and Perth. *Urban Policy and Research*, 35(1), 1–17.
Lehrer, U., & Laidley, J. (2008). Old mega-projects newly packaged? Waterfront redevelopment in Toronto. *International Journal of Urban and Regional Research*, 32(4), 786–803.
Leung, M., & Olomolaiye, P. (2010). Risk and construction stakeholder Management. In E. Chinyio, & P. Olomolaiye (Eds.), *Construction stakeholder management*. Oxford: Wiley-Blackwell.
London, K., & Siva, J. (2010). Reflexive capability maturity assessment for firm management of multi international mega projects. In *Sixth international conference on multinational joint ventures for construction works*, 22–23 September, Kyoto, Japan.
London, K., Chen, J., & Bavinton, N. (2005). Internationalisation of Australian Construction Firms, Final report, CRC_CI, Icon.Net.Pty.Ltd, Brisbane.
Lovallo, D., & Kahneman, D. (2003). Delusions of success: how optimism undermines executives' decisions. *Harvard Business Review*, 81(7), 56–63.

Mahalingam, A., & Levitt, R.E. (2007). Institutional theory as a framework for analyzing conflicts on global projects. *Journal of Construction Engineering and Management*, 133(7), 517–528.

Makui, A., Mojtahedi, M., & Mousavi, S. (2013). Project risk identification and analysis based on group decision making methodology in a fuzzy environment. *International Journal of Management Science and Engineering Management*, 5(2), 108–118.

Manowong, E., & Ogunlana, S. (2010). Strategies and tactics for managing construction stakeholders. In E. Chinyio, & P. Olomolaiye (Eds.), *Construction stakeholder management*. Oxford: Wiley-Blackwell.

March, J.G., & Olsen, J.P. (1989). *Rediscovering institutions: The organization basis of politics*. New York: Free Press.

Mawhinney, M. (2001). *International construction*. London: Blackwell Science.

Maylor, H., Geraldi, J., & Williams, T. (2011). Now, let's make it really complex (complicated): A systematic review of the complexities of projects. *International Journal of Operations & Production Management*, 31(9), 966–990.

Meyer, J.W., & Rowan, B. (1977). Institutionalized organizations: Formal structure as myth and ceremony. *American Journal of Sociology*, 83, 340–363.

Michelson, W. (1980). Long and short range criteria for housing choice and environmental behaviour. *Journal of Social Issues*, 36(3), 135–149.

Miller, R., & Hobbs, B. (2005). Governance regimes for large complex projects. *Project Management Journal*, 36(3), 42–50.

Miller, R., & Lessard, D.R. (2000). *The strategic management of large engineering projects*. Cambridge: MIT Press.

Miller, R., & Lessard, D. (2008). Evolving strategy: Risk management and the shaping of mega projects. In H. Priemus, B. Flyvbjerg, & B. van Wee (Eds.), *Decision-making on mega projects: Cost-benefit analysis, planning and innovation* (pp. 145–172). Cheltenham: Edward Elgar Publishing Limited.

Moore, C. (1986). *The mediation process: Practical strategies for resolving conflict*. San Francisco, CA: Jossey-Bass.

Morris, P., & Hough, G. (1987). *The anatomy of major projects*. London: Major Projects Association.

Mousavi, S., Jolai, F., Tavakkoli-Moghaddam, R., & Vahdani, B. (2013). A fuzzy grey model based on the compromise ranking for multi-criteria group decision making problems in manufacturing systems. *Journal of Intelligent and Fuzzy Systems*, 24(4), 819–827.

Müller, R. (2009). *Project governance*. Surrey: Gower Publishing.

Nachmias, D., & Greer, A. (1982). Governance dilemmas in an age of ambiguous authority. *Policy Science*, 14(2), 105–116.

Ninan, J., Mahalingam, A., & Clegg, S. (2019). External stakeholder management strategies and resources in megaprojects: An organizational power perspective. *Project Management Journal*, 50(6), 625–640.

Ofori, G. (1994). Formulating a long-term strategy for developing the construction industry of Singapore. *Construction Management and Economics*, 12(3), 219–231.

Olaniran, O., Love, P., Edwards, D., Olatunji, O., & Matthews, J. (2016). Chaos theory: Implications for cost overrun research in hydrocarbon megaprojects. *Journal of Construction Engineering and Management*, 143(2), 05016020.

O'Sullivan, M. (2003). The political economy of comparative corporate governance. *Review of International Political Economy*, 10(1), 23–72.

Oz, O. (2001). Sources of competitive advantage of Turkish construction companies in international markets. *Construction Management and Economics*, 19, 135–144.

Oyegoke, A. (2010). The contextual approach to stakeholder management in Finland. *Construction Stakeholder Management*, 159, 65–74.
Patel, D. (2007). Why executives should care about project governance; What your peers are doing about it. *PM World Today*, 9(4), 165–187.
Pena-Mora, F., & Tamaki, T. (2001). Effect of delivery systems on collaborative negotiations for large-scale infrastructure projects. *Journal of Management in Engineering*, 17(2), 105–121.
Peltzman, S. (1998). The economic theory of regulation after a decade of deregulation. In R. Baldwin, C. Scott, & C. Hood (Eds.), *A reader on regulation* (pp. 93–130). Oxford: Oxford University Press.
Pitsis, A., Clegg, S., Freeder, D., Sankaran, S., & Burdon, S. (2018). Megaprojects redefined – Complexity vs cost and social imperatives. *International Journal of Managing Projects in Business*, 11(1), 7–34.
Project Management Institute (PMI). (2013). *A guide to the project management body of knowledge (PMBOK Guide)* (5th ed.). Newtown Square, PA: Project Management Institute (PMI).
Porter, M. (1998). *The competitive advantage of nations* (p. 33). New York: Macmillan Business.
Priemus, H. (2008). How to improve the early stages of decision-making on MegaProjects? In H. Priemus, B. Flyvbjerg & B. van Wee (Eds.), *Decision-Making on Mega-Projects*. Edward Elgar Publishing.
Priemus, H., Flyvbjerg, B., & van Wee, B. (2008). *Decision-making on mega projects: Cost-benefit analysis, planning and innovation*. Cheltenham: Edward Elgar.
Purcell, J. (2009). Time to focus on employee voice as a prime antecedent of engagement: Rediscovering the black box. The Future of Engagement: Thought Piece Collection, www.engageforsuccess.org/futures.
Rapoport, A. (1994). *Thirty three papers in environment-behaviour research*. Newcastle upon Tyne: Urban International Press.
Remington, K., & Pollack, J. (2007). *Tools for complex projects*. Aldershot: Gower Publishing.
Rose, T., & Manley, K. (2010). Financial incentives and advanced construction procurement systems. *Project Management Journal*, 41(1), 40–50.
Ronnle, E. (2017). Planners' analysis and opportunism–benefit analysis in the swedish HSR-project: A preliminary analysis. *European Planning Studies*, 25(12), 2232–2249.
Samset, K., Berg, B., & Klakegg, O.J. (2006). Front-end governance of major public projects. Paper presented at the EURAM 2006 Conference, Oslo, May 18.
Samset, K., & Volden, G.H. (2015). Front-end definition of projects: Ten paradoxes and some reflections regarding project management and project governance. *International Journal of Project Management*, 34(2), 297–313.
Sanderson, J. (2012). Risk, uncertainty and governance in mega-projects: A critical discussion of alternative perspectives. *International Journal of Project Management*, 30, 432–443.
Sanoff, H. (2000). *Community participation methods in design and planning*. Hoboken, NJ: John Wiley and Sons.
Santoso, D.S., Ogunlana, S.O., & Minato, T. (2003). Assessment of risks in high rise building construction in Jakarta. *Engineering, Construction and Architectural Management*, 10(1), 43–55.
Sha, K. (2016). Understanding construction project governance: An inter-organisational perspective. *International Journal of Architecture, Engineering and Construction*, 5(2), 117–127.

Shenhar, A.J., & Dvir, D. (2007). *Reinventing project management: The diamond approach to successful growth and innovation.* Boston, MA: Harvard Business School Press; ISBN: 9781591398004.

Shenhar, A., & Holzmann, V. (2017). The three secrets of megaproject success: Clear strategic vision, total alignment, and adapting to complexity. *Project Management Journal*, 48(6), 29–46.

Shou Qing, W., Robert, L.K.T., Seng Kiong, T., & David, A. (1999). Political risks: Analysis of key contract clauses in China's BOT project. *Journal of Construction Engineering and Management*, 125(3), 190–197.

Siemiatycki, M. (2016). The making and impacts of a classic text in megaproject management: The case of cost overrun research. *International Journal of Project Management*, 36(2), 362–371.

Steen, J., Ford, J., & Verreynee, M. (2017). Symbols, sublimes, solutions, and problems: A garbage can model of megaprojects. *Project Management Journal*, 48(6), 117–131.

Soderlund, J., Sankaran, S., & Biesenthal, C. (2017). The past and present of megaprojects. *Project Management Journal*, 48(6), 5–16.

Steele, W., & Legacy, C. (2017). Critical urban infrastructure. *Urban Policy and Research*, 35(1), 1–6.

Thatcher, M. (2005). The third force? Independent regulatory agencies and elected politicians in Europe. *Governance*, 18, 347–374.

Turner, R., Zolin, R., & Remington, K. (2009). Monitoring the performance of complex projects from multiple perspectives over multiple time frames. In *Proceedings of the 9th International Research Network of Project Management Conference*, 11–13 October 2009, Berlin.

Turner, R., & Keegan, A. (2001). Mechanisms of governance in the project-based organization: Roles of the broker and steward. *European Management Journal*, 19(3), 254–267.

Ubaldi, B. (2013). Open government data: Towards empirical analysis of open government data initiatives, *OECD Working Papers on Public Governance*, No. 22, OECD Publishing, Paris.

van Marrewijk, A. (Ed.) (2015). *Inside megaprojects: Understanding cultural practices in project management*. Copenhagen: CBS Press.

van Marrewijk, A. (2017). The multivocality of symbols: A longitudinal study of the symbolic dimensions of the high-speed train megaproject (1995–2015). *Project Management Journal*, *48*(6), 47–59.

van Marrewijk, A. (2018). Digging for change: Change and resistance in interorganizational projects in the utilities sector. *Project Management Journal*, 49(3), 34–45.

van Marrewijk, A., Clegg, S.R., Pitsis, T., & Veenswijk, M. (2008). Managing public-private megaprojects: Paradoxes, complexity and project design. *International Journal of Project Management*, 26, 591–600.

Volden, G.H., & Andersen, B. (2018). The hierarchy of public project governance frameworks: An empirical study of principles and practices in Norwegian ministries and agencies. *International Journal of Managing Projects in Business*, 11(1), 174–197.

Wachs, M. (1989). When planners lie with numbers. *Journal of American Planning Association*, 55(4), 476–479.

Williamson, O.E. (1979). Transactioncost economics: The governance of contractual relations. *Journal of Law and Economics*, 22, 233–261.

Wilson, M. (1996). The socialization of architectural preference. *Journal of Environmental Psychology*, 16, 33–44.

Yang, J., Shen, G., Ho, M., & Drew, D. (2011). Stakeholder management in construction: An empirical study to address research gaps in previous studies. *International Journal of Project Management*, 29(7), 900–910.

Zeng, J., An, M., & Smith, N.J. (2007). Application of a fuzzy based decision-making methodology to construction project risk assessment. *International Journal of Project Management*, 25, 589–600.

Zhang, P., & London, K. (2012). 'A comparative analysis of construction industry international performance between China and United States using the International Advanced Index'. *Joint CIB International Symposium of W055, W065, W089, W118, TG76, TG78, TG81 and TG84*, Montreal, Canada.

Zidane, Y.J.T., Johansen, A., & Ekambaram, A. (2013). Megaprojects-challenges and lessons learne. *Procedia-Social and Behavioral Sciences*, 74, 349–357.

Žižek, S. (1999). *The sublime object of ideology*. Ernesto Laclau and Chantal Mouffe, London, New York: Verso.

3 Governmentality and deviance

The role of power in decision-making

What is governmentality?

Governmentality, a concept created by social thinker Michel Foucault in the 1970s, provides a helpful way to think about and understand how power works in large complex megaprojects and everyday workplaces. It helps us analyse how power is created, distributed, and used.

Governmentality is about the ways and methods of governing. It involves thinking about how people lead, manage, and control others as well as themselves. This includes the reasoning, plans, actions, tactics, choices, goals, and ways of life that are connected to leadership and management. In Foucault's examination of how governments and state agencies handle the issue of population, he observed a shift from the 16th to the 18th century. He identified two different approaches to governing, which he placed at opposite ends of a spectrum (Foucault, 1979):

- *The sovereign model*: This is modelled on Machiavellian political strategies, focusing on how a prince best maintains *power over* territory. This approach involved large, abstract, and rigid ways of thinking about power.
- *The family model*: This model is devoted to matters to enrich the small family unit.

Distinctly, Foucault identified a third approach from the mid-18th century onwards, which he termed governmentality. This approach viewed power in relation to the people it governed, considering their unique realities, characteristics, and needs. It recognised that these populations had their own dynamics, requiring government involvement even though they existed independently of government. These populations cannot simply be controlled by implementation of the law nor be thought of as a type of extended family. Foucault (1991, pp. 203–204) pointed out that governing is not merely about exerting force or compelling obedience from others. Instead, it represents a "versatile equilibrium, with complementarity and conflicts between techniques" that enable individuals to develop and transform themselves.

Governmentality – an art of governing

In other words, people are not just passive objects that the government controls, they also have their own rights and can make choices. Furthermore, each group of

DOI: 10.1201/9781003377481-3

people has unique qualities, and we need to study them through specific knowledge. This knowledge, an "art of governing", helps shape the way we govern or manage these populations. Factors relating to populations, including birth rates, longevity, demography, and public health, thus became subjects that the government studied and managed.

Foucault found an important connection between how people behave, the techniques they use to improve themselves, and how those in power encourage these techniques to achieve political goals. Of course, this doesn't mean that people have no choice or are forced to conform strictly. Instead, this type of power, referred to as "governing at a distance", offers self-improvement methods that people can adopt willingly. It helps shape individual perspectives to align with the goals of governing authorities.

Governmentality and megaproject decision-making

The governmentality approach to "governing at a distance" is well-suited for studying practices that need subtle and non-intrusive interventions, like decision-making in megaprojects (Garland, 1997). While the execution of a megaproject is typically a one-of-a-kind effort, the team members involved come from various organisations with different beliefs and ways of doing things. They often follow their usual organisational practices and routines, all while attempting to make sense of the decision-making procedures in the unique megaproject setting (van Marrewijk et al., 2008).

Hence, decision-making in megaprojects is closely connected to both official and unofficial customs, traditions, networks, and cultures. These can be easily disrupted by strict or forceful regulation. Since megaproject decision-making is part of the everyday social and economic activities within the team members' permanent organisations, any attempt to change their interests should aim to maintain the regular operations or "business as usual" (Garland, 1997). In situations marked by complexity, ambiguity, paradox, and contradiction, a suitable approach to governance is one that's subtle and non-intrusive, like governing at a distance. In this method, power is exerted by collaborating and working through intermediaries or agents. By following this approach, everyday practices can maintain their usual nature, yet at the same time achieve certain results in the context of decision-making. As described by Garland (1997, p. 187), "the situation can be governed, but it cannot be completely or coercively controlled".

Box 3.1 Example – typical ways megaproject teams work.

The following quotes from members of megaproject teams illustrate how they actually work when facing challenges on these projects:

> Informal is a lot more than formal because formal means you need to have the quorum, that kind of thing. Sometimes you don't have that

> because people are in and out. So, when there's an issue we quickly sit down and resolve it. Most important is the outcome. The outcome is the project runs smoothly.
>
> (Chief Operations Officer, Airport Operator, Airport megaproject)
>
> And that's why [when] sometimes there are certain conflicts…I can just walk up and go to AAA [the Technical Director] or BBB [the Design manager] will come down and say, "hey I need an opinion before I send this in officially, more or less is this OK or not OK?"… We may also go to them and say, "we are going to create this request do you think it's reasonable?"
>
> (Senior Manager & Head, Design & Technical, Sports Facilities Division, Sports facility megaproject)
>
> You try to run around and solicit agreement and some of these guys, they've tried to email, and we try to discourage that around here. We like to go and actually have a discussion, work it out and confirm by email ... We generally try to backdoor it and make sure that we're all on the same page and put things into the system.
>
> (Senior Design Manager, Design and Build Contractor, Sports facility megaproject)

Governmentality can serve as a useful concept to uncover and reveal how the practices of megaproject team members are structured, showing how governance frameworks are put into action in practice. As seen in the quotes in Box 3.1, megaproject clients and team members function within a paradoxical framework: ideally and rationally. Because of their training, they follow structured governance frameworks and associated roles and relationships. However, they do this while focusing on the practical aspects of getting their work done. As a result, their actual practice involves informal relationships and using different methods to deal with the challenges and uncertainties that come with megaprojects. They adapt and find solutions to make things work. The decisions made and routines and tools developed to cope effectively become the ways megaprojects are managed. These ways may or may not follow the official methods that are supposed to be used.

This view of power allows us to identify a form of governance logic – governmentality – where power functions through megaproject team members, involving how governing is done with independent individuals who have rights, rather than being imposed upon passive subjects being regulated.

A governmentality analysis seeks to study the collection of tools, which includes various practices, tactics, and strategies used to put governance mechanisms into action such as those listed in the Box 3.1 quotes. This involves examining how different individuals and participants are connected and intertwined in the processes. It looks closely at the real methods, mechanisms, steps, tools, strategies, techniques, and language used to carry out the act of governing (Boyce & Davids, 2004).

Governmentality and using power in megaproject decision-making

Decision-making in megaprojects isn't solely guided by official contracts and processes; it often results from the spontaneous use of power and other informal mechanisms (Loosemore, 1998; DeBlois et al., 2011; Ninan et al., 2019). What sets a Foucauldian governmentality perspective apart is its effort to move beyond dependence on strictly defined rules and procedures for governance. Instead, it seeks to understand power by focusing on a variety of strategic manoeuvres, tools of governance, and power structures.

Box 3.2 Lesson – relying on rationality and logic without considering the influence of other types of power is unlikely to result in successful outcomes.

According to a case study by Flyvbjerg (1998) of a complex large-scale urban project in Denmark known as the Aalborg project, stakeholders' actions were strongly influenced by the covert exercise of power and the protection of special interests. A key lesson which can be learnt from the case study is that "power defines what counts as rationality and knowledge and thereby what counts as reality" (Flyvbjerg, 1998, p. 319). In other words, political power can sometimes override rational decision-making and lead to dominance. However, this doesn't mean that megaproject team members should disregard rational arguments based on factors like the environment, society, ethics, and professionalism. Instead, relying solely on rationality and logic, while ignoring other forms of power, is unlikely to lead to successful results.

Foucault's concept of governmentality shows that power is mainly concerned with influencing or shaping behaviour intentionally and purposefully to achieve a variety of ends, much like what we see in the Aalborg case study. This perspective views power as a form of guidance and acknowledges that there are multiple approaches to decision-making. It suggests that strict and rule-bound structures, which are common in current megaproject governance, may not be the only way of governing. This doesn't imply a complete departure from traditional sovereign models but suggests that this form of governance coexists with other forms of governing practices.

What the Aalborg case study also shows is that the capacity to use power is like a skill or craft. It involves clever planning, creativity, tacit skills, and practical know-how, among other things. Therefore, it is crucial to grasp how project team

Table 3.1 A governmentality approach focuses on a combination of governance and governing in the use of power in megaproject decision-making

	Governance	*Governing*
View of the decision-making environment	Idealistic	Realistic
Means	Prescribed standards	Practised "norms"
Mode	Formal legitimate authority	Informal use of power
Level of analysis	Macro	Micro

members and stakeholders use power when making decisions on megaprojects. This means looking beyond "what is said" and considering their actions or "what is done".

The focus of governmentality work is not on creating an ideal model but on creating a practical map that shows the many different ways governance mechanisms are put into action in real-world megaprojects by various authorities. It looks at both the big picture and the day-to-day practices, including the systems and routines that are often assumed to be rational and normal, as detailed in Table 3.1.

Megaproject decision-making: a collective effort

Megaproject decision-making, in this sense, involves a collective effort. It encompasses a variety of governing bodies and authorities, different aspects of behaviour to regulate, diverse objectives pursued, various norms referenced, and a range of effects and consequences. The art of governing lies in maintaining a delicate balance between consensus and the use of force.

Box 3.3 Lesson – effective decision-making in megaprojects involves finding the right balance between achieving consensus and exercising authority when necessary.

The following quote illustrates how important it is to skilfully balance between consensus and the judicious use of authority:

> Human factor with integration is very important. Even if you have a good contract, you have targets, measurement tools and so on. At the end of the day, you are dealing with humans and so its human factors are important
>
> (Chief Executive Officer, Special Purpose Vehicle, Sports facility megaproject)

Several important questions to consider include the following:

How are megaprojects shaped within the network of power relations?
How are ideas or propositions disseminated, accepted, or rejected?
What are some of the accepted norms of behaviour or "rules of the game"?
What forms of power are available to the various "authorities"?
What are the different ways in which clients and project team members accommodate, create, and nurture power?
When is it deemed appropriate to practice deviance to influence decisions?
What are the factors that influence how clients and project team members navigate decision-making through deviant behaviour?
What conditions promote constructive or destructive deviance?

The general approach undertaken to understand and improve megaproject performance has often concentrated on achieving compliance through the transfer of political power from those at the highest level to those in less powerful positions. This approach, even though grounded in sound principles, is just one way of governing.

The real story of megaproject decision-making is complex and full of compromises, adapting to different needs, uncertainty, and contradictions, and sometimes only partially carrying out planned policies. To gain a better understanding of the megaproject decision-making environment, it's valuable to examine the subtle and everyday use of power that might not be immediately apparent.

Definitions of power

The word "power" is frequently used, and people often assume its meaning without providing a clear definition (Galbraith, 1983). An early definition was provided by Weber (1947) as the capacity of an individual within a social relationship to achieve their own objectives or enforce their own will, even in the face of opposition or resistance from others.

In other words, power means the ability to get others to agree with one's choices, influence, or control how they act, or more informally to "get one's way" (Brislin, 1991, p. 2; French & Raven, 1959; Fiske & Depret, 1996; Turner, 2005).

Power can be manifested and employed in various contexts and domains (Carroll & Buchholtz, 2006):

- Economic
- Political
- Social/cultural
- Technological
- Environmental power

In this sense, power "belongs to things as well as people" (Turner, 2005, p. 6). By drawing upon a common pool of diverse resources, people can intentionally achieve the outcomes they have planned in a social transaction (Wrong, 1979; Ewick & Silbey, 2003). Power is, however, not a possession. Building on Weber's definition, power is a social relationship where the outcomes depend on what all the people involved do (Ewick & Silbey, 2003).

Power and megaproject decision-making

Power is central to human affairs and is part of our everyday and professional lives (Turner, 2005). While it is not the only factor, it certainly plays a central role in decision-making and in how megaprojects are shaped.

The megaproject decision-making environment consists of networks of power dynamics involving a diverse range of stakeholders such as the client, project manager, architects, contractors, users, property owners, financiers, regulatory bodies, and the local community. Each of these stakeholders contributes their own perspective and importance at different project stages. In megaprojects, there are fundamental tensions between stakeholders due to their competing values and interests. This leads to an unequal distribution of power (Wild, 2002, Ninan et al., 2019).

Such a complex setting, marked by unequal power distribution and conflicting institutional logic, makes it challenging to apply standard project management routines for decision-making (Ninan et al., 2019; Biesenthal et al., 2018). Some stakeholders, even if they are not directly involved with the planning and building of megaprojects, are still impacted by the project's outcomes, which can affect their future or daily lives.

Megaproject decision-making can be compared to networks of self-interested stakeholders who must find ways to work together despite their differences. As described by Sayer (2001), networks do not invariably blend the self-interests of diverse actors into a cohesive and equal entity; instead, they may exhibit imbalances in power, strategic alliances, deceptive practices, and opportunistic cooperation. What might seem to signify trust could, in reality, be predominantly a result of dominance, a dearth of alternatives, or mere mutual dependence (Sayer, 2001).

Effective megaproject decision-making is carried out by clients and team members who understand the importance of power and view it as a means to achieve their desired outcomes, essentially knowing how to "get things done" (Brislin, 1991). Highly skilled clients and team members recognise the importance of power in decision-making and in the allocation of scarce resources on megaprojects. They also view power holders in terms of their potential influence or clout in helping them achieve their desired outcomes. For these clients and team members, the focus is on the constructive aspects of power, and the primary goal is to "get things done", rather than dwelling on the negative aspects of power (Brislin, 1991).

Models and types of power

Power has often been cast in negative terms and has rarely been seen as a way to enable and harness creativity positively. Karlberg (2004) identifies four power distribution models, with three of them having a negative view of power:

- *Power over*: To have power *to* and to have power *over* (culture of a contest) results in a state of inequality between parties coupled with adversarial relationships, viewing humans as aggressive and greedy by nature, hence "natural selection".

- *Assisted power*: This results in inequality but might help develop mutual relationships.
- *Balance of power*: This results in equality between parties with the probability of the parties resorting to adversarial attitudes.
- *Mutual empowerment or power with (mutualism)*: Advances relationships on an equal basis.

Positional power

The orthodox view of positional power sees power as a form of command and control over scarce resources, often leading to inequality or adversarial attitudes. Positional power is based on hierarchy and structure, where authority is gained through one's position within an organisation or network. This type of power is often viewed as a fixed and finite resource, leading to competition among actors.

Within the context of megaprojects, governance frameworks are typically implemented based on the formal authority of those with positional power over project teams, using compliance, coercion, and strict control. Project teams are expected to resolve project issues and meet schedules and budgets under very rigid conditions. This approach can be problematic because standard project management routines would likely be inefficient for addressing complex challenges that megaprojects often present.

Box 3.4 Governing methods using positional power.

Foucault gives examples of various methods of governing using positional power. He points out that the Christian pastorate serves as a spiritual way to govern the soul, while reason is used as a political means to govern the population. These "governmental technologies" help establish and control power dynamics, which can sometimes result in dominance (Lemke, 2000).

Foucault's idea of government technologies goes beyond the mere spontaneous use of power. It refers to the structured management of behaviour by applying suitable technical methods in a more or less rational way.

The formal governance methods using positional power aim to regulate the behaviour of project team members. They are founded on the idea that these governmental technologies provide unbiased and just guidance. According to this perspective, following the established procedures and protocols is more important than the impromptu decisions and actions taken by team members. The focus is on adhering to rules and technical instructions, where team members are viewed as passive entities merely performing their roles and interacting with others in an almost "thing-like" way. Any behaviour that deviates from the prescribed procedures and protocols is considered unacceptable and may lead to some form of punishment or penalty.

Box 3.5 Formal megaproject governance mechanisms.

In megaprojects, these governmental technologies encompass the structured governance methods to ensure contractual oversight and, consequently, compliance. This involves techniques like process analysis, cost–benefit analysis, progress reviews, and document control systems. The goal is to establish a tightly regulated environment with clear accountability for decision-making, as well as well-defined roles and boundaries for different parties. The formal structures and protocols aim to create a transparent trail of official document exchange among project stakeholders.

Textualisation, which is the use of written contracts, progress reports, and record keeping, plays a central role in these governmental technologies. Texts are highly authoritative because they can be saved, retrieved, and reviewed (Ewick & Silbey, 1998). According to this perspective, it seems that tightly controlled procedures and protocols take the place of decisions undertaken by project team members:

> It's the contract that manages the job.
>
> (Chief Operating Officer, Special Purpose Vehicle, Sports facility megaproject)

> If we hear something that is different from what has been prescribed in the agreement, we'll pull everybody back…so they don't run off and do their own things.
>
> (Deputy Director, Operations & Programming, Sports Facilities Division, Sports facility megaproject)

> So how this matrix works…I have a specific person who shall I interact with…I will go to my superior and similarly my down liner, they understand their level. They are not supposed to just do anything
>
> (Senior Construction Manager, Airport Operator, Airport megaproject)

Relational power

In contrast, relational power is based on a more dynamic idea of "power with" and "power to" which extends the concept of "power over" to one which is co-active in nature (Follett, 1940; Haugaard, 2012). According to Foucault, power can have both constraining and productive aspects and should be understood in terms of strategies, discourses, and processes rather than positions and resources (Clegg, 1989, 2014). A relational view of power is built on the concept of growth capacity, where both constraints and enabling forces coexist, allowing individuals to see and express things they couldn't before.

Based on this relational perspective, power is not a finite resource that can be distributed or balanced amongst competing needs. Central to this view of relational power is the idea of "circular response", where the interaction between subjects and objects is dynamic and interdependent. While ideas of positional power are associated with linear assumptions, the concept of power with and power to is linked to a more dynamic, interrelated, and process-based perspective.

Although power relations often involve asymmetry, where those with greater access to resources can exert more control over others, it doesn't mean that the dynamics or established power relationships cannot be challenged or changed. Various terms like secondary adjustments (Goffman, 1961), tactics (de Certeau, 1984), weapons of the weak (Scott, 1985), or acts of resistance (Ewick & Silbey, 2003) describe the everyday strategic approaches through which people in relatively powerless positions adapt to power while safeguarding their interests (Ewick & Silbey, 2003).

Relational power empowers individuals to be more than just objects of regulation; it allows them to become active subjects with rights. This particular understanding of power is associated with the concept of governmentality.

Box 3.6 Governing methods using relational power.

The following quotes from members of megaproject teams illustrate the different ways they use relational power in their everyday dealings with others on projects:

> Our meetings basically – consultant meetings - so we just say, "let's have a meeting" and then we just meet. That happens a lot.
>
> (Chief Operations Officer, Special Purpose Vehicle, Sports facility megaproject)

> When you are one of the bosses or one of the chief or part of the army you must know who is your lieutenant. Your people who help you and I know the people who help me, and I know also what I can get from this guy and also the way to get it. So yes, you need to know your troops and the leaders of your troops. And you know, they are not leading their troops the same way, one to the other, you need to know how they function…Yes, I put things in place to help them.
>
> (Technical Director, Design and Build Contractor, Sports facility megaproject)

Power as guidance

While the word government may imply a strictly political meaning today, Foucault took a broader perspective that included philosophical, religious, medical, and familial contexts (Lemke, 2000). He defined government as "the conduct of conduct", meaning it is a type of activity aimed at influencing or guiding the actions of others using various techniques of control and self-regulation (Gordon, 1991). Foucault's analyses are not limited to "the state"; they extend the idea of "governmental authorities" to include families, churches, experts, professions, and all the different powers involved in shaping behaviour. This blurs the lines between "public" and "private" or between "state" and "civil society" (Garland, 1997).

Additionally, defining government as the "conduct of conduct" invites exploration of self-government. This extends beyond how one exercises power over others to how one governs their own behaviour (Dean, 2010). "Conduct" within the context of governmentality has a dual meaning and plays on different aspects of the word "conduct" (Dean, 2010, p. 17):

- *To lead, direct, or guide*: This implies that some form of calculation, manipulation, or strategic thinking is involved in figuring out how to do this.
- *To behave*: This refers to individual sets of behaviour and actions. It suggests that individuals are expected to work towards self-guidance or self-regulation based on normative conduct serving as an ideal. Furthermore, there are also agents responsible for ensuring such regulation takes place.

By considering the word "conduct" in its various meanings, we can define government as any effort to influence behaviour based on specific norms, often with some degree of intent and purpose, aiming to achieve different objectives. In this context, power primarily involves providing guidance, essentially "governing the forms of self-government" and shaping the range of actions available to individuals (Lemke, 2000, p. 3).

The circulation of power

The concept of governmentality enhances our comprehension of power by showing that it isn't confined to a single entity or institution, like the state. Instead, power is found in various places, involving a range of "authorities" engaging in governmental practices (Garland, 1997; Rose et al., 2009). Foucault encourages an examination of power that sees it as continually circulating, not just as a commodity to be owned by someone, but as something pervasive and flowing like a chain.

In this perspective, power is not a static possession or an attribute that one can hold or acquire like wealth or property. It operates within a network or a net-like organisation. Moreover, people are not just passive recipients; they both experience and contribute to the flow and function of power (Foucault, 1980).

Foucault talked about the constant flow of power and called it omnipresent, saying it comes from everywhere (Foucault, 1980). Scholars who have adopted Foucault's idea of power circulating recognise that power can be seen in everyday life and institutions (Rose & Miller, 1992; Foucault, 1980; Donzelot, 1979; Rose et al., 2009).

Box 3.7 Case study – power circulating in the realm of "the social".

An example of how power is seen as circulating is shown by Donzelot, who looked at how public authorities, like doctors, social workers, and psychiatrists, get involved in family matters. He showed that various forms of authority exist between the government and individuals. Donzelot explained how government policies and the practices of professions like social work, mass education, psychiatry, and philanthropy changed family life and made regular family members act on behalf of the government.

Donzelot called this intricate network of governance "the social", made up of government agents spread across hospitals, schools, social work offices, juvenile courts, and clinics. This is where "governmental power flows through and acquires its effectiveness from this network of professional enclosures and agencies" (Garland, 1997). According to this perspective, power relations are deeply rooted in the realm of the social, working in the space between government and individuals (Garland, 1997).

It's important not to assume that just because a structure exists within a network, as often seen in large organisations, it automatically means that members accept or implement it. While these formalised structures may seem very visible, there are also other "spheres of authority" with less obvious relationships and interactions where power is constantly at play. In these cases, formalised structures might play a strategic role, but they are not necessarily the dominant force (Sending & Neumann, 2006).

The circulation of power in megaprojects

Megaproject governance often involves strict contractual enforcement with extensive monitoring to ensure compliance (Clegg et al., 2002). Various governance tools are used to control outcomes, such as process analysis and cost–benefit analysis. A common criticism of existing governance approaches, primarily based on the sovereign or discipline model, is their narrow focus on technical and quantitative decision-making with conformity as the main goal. This "decision-making by numbers" method assigns numerical values to specific activities while downplaying the importance of less tangible or unquantifiable factors where power circulates (Boyce & Davids, 2004).

Despite aiming to minimise conflicts and differences among multiple parties, contractual agreements frequently become sources of disputes because complex contract documents are open to interpretations, and project team members often have conflicting interests (Clegg et al., 2002). The governmentality perspective suggests that numerous "authorities" govern in different locations, each with competing interests (Garland, 1997; Rose et al., 2009).

Box 3.8 Lesson – it is important to recognise that a diverse group of stakeholders can exert influence and act as authorities, impacting the direction of the project.

A variety of indirect, informal, or non-expert participants can sometimes act as authorities on megaprojects, influencing project directions. These authorities operate in various sites, far removed from the high-level governance mechanisms primarily concerned with contractual oversight. They are also referred to as "stakeholders of the shadows" (Winch, 2017). These authorities act as another layer of "client" that might not be bound by the formal project team's contractual scope, budget, or timelines, and thus, they may wield power without having corresponding responsibilities (Ninan et al., 2019).

A governmentality study of the megaproject decision-making environment aims to reveal how governance frameworks and mechanisms actually impact the practices and processes of these diverse authorities in various locations where power relations are played out.

Diagrams of power

Foucault described power as a system of relationships among different forces, and he represented these relationships using what he called "diagrams of power" in a topological way.

Box 3.9 Case study – diagrams of power

Munro (2000) conducted a study on power relations related to advances in information technology and outlined three key diagrams of power: the sovereign, panopticon, and network diagrams. Table 3.2 summarises the structures, key characteristics, and distinctions between these three diagrams of power. The first two, the sovereign and panopticon diagrams, were initially identified by Foucault during his examination of the prison system, while the network model was developed by Munro (2000) in his exploration of information technology developments.

Munro's (2000) concept of a network diagram, which represents power relations as interconnected nodes, offers a more realistic depiction of how power influences decision-making in megaprojects. The argument put forth by Munro (2000, p. 188) is not that network power has replaced disciplinary or sovereign power, but that it is currently reshaping social relations and enabling the application of various forms of power. In fact, we can observe the continued presence of all these power forms in various aspects of society today. Disciplinary power remains evident in our schools, factories, and the growth of the prison population, while sovereign power continues to hold sway even within the context of the networked society.

Table 3.2 The three diagrams of power

Diagrams	*Sovereign*	*Panopticon*	*Network*
Topology	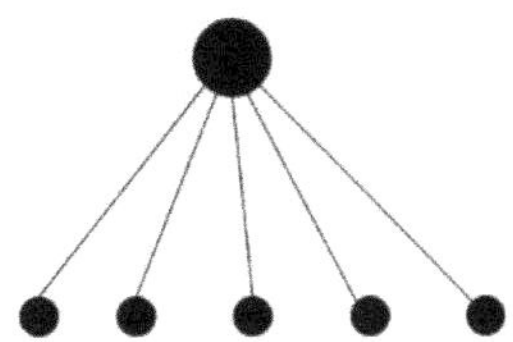	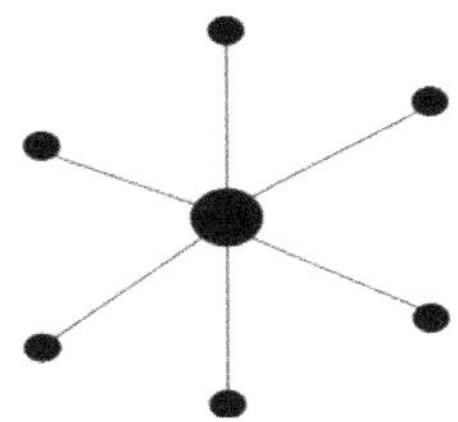	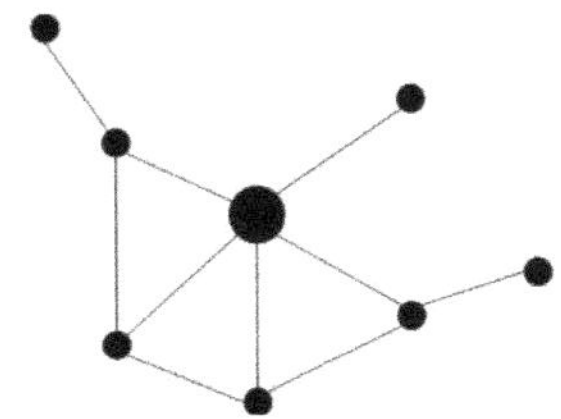
Ontology of power	Power over	Power over	Power with, power to
Technique	Power radiates from the sovereign down upon their subjects. Prison was a marginal form of punishment. Torturous spectacles such as burning and quartering were the order of the day, which openly displayed the power of the regent	Those confined on the periphery of the Panopticon are cut off from each other, yet are constantly under surveillance by the central column. The prison became the primary instrument of punishment for criminals	Made up of a series of interconnected nodes, where potentially any point in the network can communicate with any other point. It should also be noted that other relationships might exist within this diagram, such as Panoptic or sovereign relations
Mechanism of power	Disciplinary power	Disciplinary power	Non-disciplinary power

Adapted from Munro (2000).

Munro's (2000) research primarily examined how contemporary advancements in information technology are reshaping power relationships in modern society. However, the insights gained from this study can be applied to gain a deeper understanding of decision-making in the megaproject environment within the construction industry. In the context of megaprojects, the current governance mechanisms designed to ensure adherence to contracts share similarities with the sovereign and panopticon forms of exercising power. The argument is that the prevailing focus on governance mechanisms in megaprojects, which is often based on disciplinary approaches, may not fully address the complexities of power dynamics within megaproject coalitions.

Diagrams of power in megaprojects

A similar observation can be extended to megaprojects, which are known for their intricate power dynamics, political intricacies, and conflicting interests (Marshall, 2006; Pinto, 1996; van Marrewijk, 2015). In a recent study (Ninan et al., 2019) that explored strategies for managing external stakeholders in megaprojects, it was found that a range of approaches was used, such as persuasion, deputation, and give and take. While the study did not explicitly align megaproject decision-making practices with the three diagrams of power, it seems that the strategies identified bear some resemblance to the techniques described in the sovereign, panopticon, and network models.

A Foucauldian governmentality study offers a valuable approach for analysing megaproject decision-making as a field of power relations that consists of complex networks of authorities using various diagrams of power. In this perspective, megaproject decision-making is characterised by a governmentality framework that involves multiple diagrams of power, blending diverse governance mechanisms with various governing practices. This interplay of different modes of governing creates sophisticated policies, practices, and processes, which cannot be simplified into a single formula (Garland, 1997). Therefore, it's crucial to consider not only the examination of grand techniques like the Panopticon but also the everyday government techniques and tools.

When conceptualised through governmentality, it becomes evident that this perspective can offer insights into how various forms of power function through consent rather than coercion. Megaproject team members often engage in strategic interactions or games as they work to address project challenges, and this approach helps us understand how power dynamics are at play in these situations. Instead of relying solely on force, power can operate through the consent and strategic actions of those involved in megaproject decision-making.

Megaproject decision-making and strategic games of power

Strategic games, or games of power, are a fundamental aspect of human interaction and can be observed in various contexts or spheres (Lemke, 2000). These games of power can take many forms, including rational arguments and strategic manipulation.

However, it's important to note that the exercise of power in this way doesn't always entail the imposition of one's upon another party. Moreover, using power in such a manner doesn't necessarily lead to a loss of personal liberty. On the contrary, it can result in a kind of "empowerment" or "responsibilisation" of individuals, compelling them to make independent decisions within the range of possible actions (Lemke, 2000, p. 5).

Box 3.10 Working out deals through strategic games of power.

Warner (2010) suggests that individuals in less powerful positions can strategically negotiate and create connections to enhance their standing, compensating for power disparities in one area with strengths in another. When it comes to project team members seeking to influence decision-making in megaprojects, it is expected that they would aim to build connections or relationships with those occupying central positions in the power structure to strengthen their own position.

Conversely, those in more advantageous positions might also be interested in negotiating and forming agreements to achieve their objectives. As Warner (2010) points out, "even the powerful are ultimately interested in some levelling of the playing field to arrive at a constructive, durable result". This emphasises the idea that both weaker and stronger parties can benefit from working out deals and fostering cooperation in the pursuit of their goals.

Strategic games operate on the premise that megaproject decision-making resembles a game with rules that can be stretched or exploited to one's advantage. In contrast, a more idealised perspective of governance sees policies and guidelines as limiting and sometimes pre-empting human action. The game-like view sees rules as creating spaces, often referred to as "loopholes", within which deviant behaviour can occur and advantage taken. Rules, in this context, both set limits and offer opportunities for participants in terms of their actions.

While rules provide a degree of predictability in shaping behaviour, they can also be circumvented. In other words, even though formalised governance policies and procedures offer rationalised instructions and guidance, they don't entirely determine how individuals will actually act and behave. Actors within megaproject decision-making have room for manoeuvre and interpretation within the framework of established rules and procedures.

Exercising power through deviance

Megaprojects often experience deviations due to their inherent complexity, substantial costs, and ambitious goals (Morris & Hough, 1987; Fiori & Kovaka, 2005; Pinto, 2014). Deviance, in this context, refers to actions or events that diverge from established norms, standards, or rules (Warren, 2003). These events, which can disrupt project plans in terms of scope, time, or cost, can vary in severity, ranging from minor issues or tensions to significant crises (Hallgren & Soderholm, 2010; Loosemore, 1998). Regardless of their scale, they are all unexpected events that stand out from normal operations because they demand the attention of the project team due to their impact on scope, time, or cost objectives (Hallgren & Soderholm, 2010).

Research on deviant behaviour can be broadly categorised into two groups: constructive and destructive deviance. Constructive deviance examines the positive effects of deviant behaviour, while destructive deviance focuses on its negative consequences. Therefore, the literature on deviance suggests that such behaviour can be linked to both desirable and undesirable outcomes (Warren, 2003; Bodankin & Tziner, 2009).

Box 3.11 Destructive deviance and constructive deviance.

Destructive deviance is characterised as deliberate and voluntary behaviour that violates significant organisational norms, posing a threat to the wellbeing of the organisation, its members, or both (Robinson & Bennett, 1995; Bodankin & Tziner, 2009). These departures from established norms have been demonstrated to lead to disastrous consequences, affecting not only organisations but entire industries and societies (Vaughan, 1996; Warren, 2003).

Destructive deviance can be categorised into two groups (Bodankin & Tziner, 2009):

- *Interpersonal destructive deviance*: This category involves actions directed at individuals. Examples include behaviours such as stealing from other employees and informing them.
- *Organisational destructive deviance*: This category focuses on actions that are directed at the organisation itself. Examples include acts like embezzlement from the organisation or deliberate sabotage of equipment and resources.

Constructive deviance is described as voluntary and intentional behaviour that violates significant organisational norms in honourable ways, leading to the wellbeing of the organisation, its members, or both (Bodankin & Tziner, 2009). In other words, while it involves breaking organisational norms, this behaviour can also be seen as beneficial because it assists the organisation achieve its objectives in some form (Bodankin & Tziner, 2009). Constructive deviance can foster innovation and creativity, thereby contributing to an organisation's competitive advantage and overall wellbeing. It's important to note that while adherence to prescribed norms is crucial for the smooth operation of organisations and projects, strict compliance with rules can sometimes hinder employees from discovering innovative solutions to problems.

Constructive deviance can also be categorised into two groups (Bodankin & Tziner, 2009):

- *Interpersonal constructive deviance*: This involves actions directed at individuals. An example is disobeying managerial orders to improve organisational processes.
- *Organisational constructive deviance*: This includes actions directed at the organisation as a whole. Examples include finding innovative or creative ways to solve problems and challenging or breaking rules to address client problems.

Broadly speaking, constructive deviance is an umbrella term encompassing various behaviours, including principled organisational dissent, whistleblowing, organisational citizenship behaviour, and functional or creative disobedience.

Normalisation of deviance

The term "normalisation of deviance" was first introduced by Vaughan (1996) in her research on the culture of National Aeronautics and Space Administration (NASA) before the Challenger disaster. Vaughan's work revealed that people tend to become more accepting of deviant behaviour the more they are exposed to it. This normalisation occurs when individuals and organisations consistently deviate from established standards without facing consequences, leading to a situation where these violations are regarded as the form and the usual way of doing things (Hajikazemi et al., 2020).

In essence, the normalisation of deviance occurs when deviations from established norms or standards become so frequent and accepted that they are no longer seen as exceptional or problematic. This phenomenon can be particularly dangerous in contexts where adherence to rules and standards is critical, as it can lead to complacency regarding deviant behaviour and a failure to address the underlying issues that contribute to it.

The concept of normalisation of deviance highlights the role of "gradualism" in fostering tolerance for deviant behaviour (Pinto, 2014). Over time, individuals and organisations become acclimatised to deviating from established norms when it appears that these deviations do not result in any apparent negative consequences (Hajikazemi et al., 2020). Gradualism involves making incremental adjustments to what is initially accepted as the form, with these adjustments becoming increasingly accepted until they become the new standard.

As pointed out by Pinto (2014, p. 377), the normalisation of deviance implies that what was initially considered unexpected eventually becomes expected, and what is expected eventually becomes accepted. This process reflects how deviations from established standards can become ingrained in the culture and practices of a group or organisation over time, often with significant consequences.

Box 3.12 Lesson – identifying and interrupting the normalisation of deviance is critical for project success.

The concept of gradualism and the normalisation of deviance have been explored in the context of project management, revealing their impact on project costs and schedules. Past studies have shown how incremental changes to project scope and modifications to plans and control documentation can lead to the escalation of project costs and timelines (Pinto, 2014).

More recent research (Hajikezemi et al., 2020) has shed light on how the gradual process of normalising deviant behaviour can be influenced by actors both within and external to an organisation. These findings highlight the importance of recognising and addressing destructive deviance to mitigate the risk of undesirable consequences for organisations.

In summary, identifying and interrupting the normalisation of deviance is essential to maintain project and organisational integrity, prevent cost overruns, and ensure that projects are completed successfully and within the established parameters.

Megaproject deviance

Defining the concepts of constructive and destructive deviance within the context of megaproject practice is essential for understanding and managing these behaviours effectively. In the context of megaprojects, project contracts serve as the formal guidelines for the project. They establish the organisational structure, lines of communication, and the process of exchanging documents for coordination and decision-making. Therefore, understanding how constructive and destructive deviance can impact these formal project agreements and processes is crucial for ensuring successful project outcomes.

Constructive deviance includes behaviours that violate or depart from the prescribed standards of the megaproject but have a positive impact on the project. An example of this behaviour is when a project team member breaks the rules to address a project issue, ultimately advancing the project with minimal time loss. This behaviour is often characterised by an altruistic, collaborative, or ethical nature.

Destructive deviance, on the other hand, encompasses behaviours that violate or depart from the megaproject prescribed standards but have a negative effect on project outcomes. An example of destructive deviance is when a project team member intentionally hinders project activities to benefit their own organisation, even if it harms the megaproject, potentially leading to delays and cost overruns. This behaviour is often associated with opportunistic and deceptive characteristics.

Box 3.13 Lesson – not all deviations are destructive; some can be constructive or even necessary.

The recent research undertaken by Chan and Jia (2019), which investigated the implementation of intended policies and procedures in a megaproject, found that deviations were not the result of deceitful behaviour. Instead, these deviations emerged as a response to the challenges of adhering to governance mechanisms while dealing with the inevitable paradoxes arising from the changing conditions of the project. This highlights the complexity of megaprojects and the need to explore various forms of deviant behaviour that may occur within this context. Understanding the nature and motivations behind these deviations is crucial for effectively managing megaprojects.

The research on megaprojects highlights that despite the importance of formal and organisational and project structures in establishing communication flows, project coordination and decision-making, the way work is carried out daily is often more influenced by informal relationships and interactions among team members, both within and across organisations (Clegg et al., 2002; Siva & Gajendran, 2023).

Formal mechanisms, which are overly prescriptive and highly structured, can sometimes appear detached from the actual concerns and realities faced by project team members. This misalignment can render governmental technologies less effective in addressing the complex challenges of megaprojects. As a result, project team members often resort to "muddling through" and seeking alternative solutions that may not conform to existing governance mechanisms (Ninan et al., 2019).

Examining these hidden everyday deviant acts can shed light on why governance mechanisms may not be fully embraced in megaprojects. It allows us to understand how and why the specific interests of specific team members or authorities are advanced while others may not benefit to the same extent. This insight is crucial for improving the effectiveness of governance in megaprojects and ensuring that it aligns more closely with the practical needs and concerns of those involved.

Everyday deviant tactics

Governmentality theory makes a clear distinction between power and domination. Domination, in this context, refers to asymmetrical power relationships which are stable and hierarchical, making them challenging to reverse (Lemke, 2000). Domination leads to those in less powerful positions having limited room for manoeuvre. Faced with such limited capacity, individuals in subordinate positions may find themselves compelled to resist or rebel against the governmental technologies they are meant to adhere to. In the absence of alternatives, those in subordinate positions may "exploit the interstices of conventional social practices to forge moments of respite from the power of law" (Ewick & Silbey, 2003) and do so by employing various deviant tactics.

Box 3.14 Exercising power through everyday deviant tactics.

There are various ways in which megaproject team members can exercise power through deviant tactics:

- *Rule literalness*: This tactic relies on the premise that all transactions are governed by rules. Incomplete or vague systems offer opportunities for individuals to exploit, either to achieve compliance or resistance.
- *Masquerade*: This involves manipulating roles to influence transactions and often involves some degree of deception. An example of a masquerade is presenting oneself as less experienced or knowledgeable to avoid work.
- *Disrupting hierarchy*: Individuals who purposely ignore hierarchies of systems or processes and the associated lines of authority, respect, and duty. By skipping levels of hierarchy, individuals can achieve their objectives. A related term is "issue selling", which involves using voluntary behaviours to influence the organisational agenda by getting those above them to pay attention to an issue.
- *Friendship and social ties*: This relies on emotional relationships with others. Deals are often made because certain people have warm ties with one another. People are more willing to comply with the requests of friends or those they have warm social ties with, making it easier to gain their support.
- *Rational persuasion*: This involves persuasion using specialised information or logical arguments, typically in the form of expert advice. This tactic places certain individuals in positions of legitimacy or authority. It is subject to competition from other experts and disputes.
- *Scarcity*: Individuals use their access to scarce goods, such as networks or expertise, to exert power. These individuals may possess specific scarce skills, knowledge, or expertise, making them powerful figures who can choose whom they provide their service to.
- *Legitimacy*: This form of power is derived from formal positions within an organisation or network, granting its holder automatic rights to access, organise, and have power and influence over information.
- *Foot dragging*: This is a form of resistance that depends on and defies the rationalisation of time.
- *Gatekeeping*: This involves the control of flows of information, including the deliberate withholding or filtering of information, as a way to exercise power over others.
- *Swapping or exchanging deals*: This refers to the exchange of favours or a willingness to reciprocate. It is based on the promise of access to shared benefits.
- *Pressure:* This involves the use of demands, threats, or persistent reminders to influence others.

However, it is important to note that deviant behaviour on megaprojects may be driven by a variety of factors, and not solely as a response to domination. The motivations and contexts for deviance can be multifaceted and may include factors related to project complexities, conflicting interests, or individual perspectives, among others.

The identification of various deviant tactics in the literature provides valuable insights into how individuals attempt to influence and shape power relations to achieve their desired outcomes. These deviant tactics are essentially everyday strategies employed to manipulate, challenge, or navigate power dynamics, and have been drawn from social psychology, organisational psychology and management, and political and legal theory. While these tactics may originate from diverse disciplines, their underlying principles can serve as a useful foundation for understanding the intricate dynamics of megaproject practice.

It is important to recognise that these tactics are likely to be used in combination, and they may be deployed both openly and covertly.

Hidden power

The concept of governmentality emphasises a departure from grand theory and a focus on the "grey sciences" and the "minor professions" that deal with the everyday governance of economic and social life. It places a spotlight on the mundane aspects of governing daily life and recognises a less obvious form of power that often goes unnoticed.

Foucault's perspective on the study of resistance to power provides a valuable starting point for understanding the "economy of power relations". This approach aims to reveal power relations, identify their positions, understand their points of application, and explore the methods employed. According to Foucault, one should strive to locate power at the extreme points of its exercise, where it tends to be less legal in character.

Box 3.15 Case study – using hidden power to "make things work"

Garland's (1997) investigation of crime control from a governmentality perspective revealed a dimension that is not purely theoretical or based on official policy statements. Instead, it delves into the practical aspects of governance, focusing on problem-solving and the effort to "make things work". He terms this approach "the new criminologies of everyday life".

In this view, crime is seen as a routine phenomenon rather than a disruption of normality. To understand it fully, one must pay careful attention to normal, mundane, and often hidden events and activities. This perspective emphasises the importance of examining the practical, day-to-day aspects of governance, and power relations in various domains, including crime control, which may not be evident when looking solely at policy statements and official rhetoric.

Expanding on the idea of resistance to power as being "less legal in character", Foucault explores the emergence of disciplinary administration as the "dark side" of law. This suggests that the exercise of power can often take forms that are not strictly in line with legal or officially sanctioned methods. It underscores the importance of examining power dynamics in their various manifestations, even those that might not conform to traditional legal structures.

The study of power should encompass not only explicit and formal macro power associated with government technologies but also the "dark side" of power that resides in dispersed sites of domination, consisting of what Foucault referred to as "tiny, everyday…systems of micropower". These micropower dynamics are often "less legal" in character and are enacted by individuals who may lack the capacity to openly challenge or contest power. Consequently, these practices are intentionally designed and executed to remain hidden and undetected by those they target. Because of their ubiquitous nature, these daily deviant and rebellious acts are not easily identifiable.

Within the context of megaprojects, these everyday deviant tactics, which are rooted in covert power, may include masquerade or deceit, opportunistic behaviour, gatekeeping, and foot dragging. It is essential to recognise and understand these less obvious forms of power and resistance, as they play a significant role in shaping the dynamics of megaproject decision-making and governance.

Using covert power in megaprojects

The distinction between overt and covert dimensions of power is a valuable framework for explaining the concealed and everyday practices not only in crime control within megaprojects but also in routine decision-making. Overt power is readily observable as it involves the direct exercise of control over others. In contrast, covert power is a hidden form of power, often concealed within enduring institutional practices and taken-for-granted ideas (Clegg, 1989).

These concealed practices and ideas are often referred to as "hidden structures of power" or "deep structures" because they are tacitly accepted. As a result, they continue to be reproduced (Ninan et al., 2019; Clegg, 1975). Although subtle, the exercise of covert power can be highly effective in shaping decision-making processes and governance within megaprojects. This distinction helps shed light on the less obvious but influential dynamics that operate within these complex endeavours.

Box 3.16 Case study – keeping illicit activities hidden.

Van de Bunt's (2010) investigation of major fraud scandals highlights an important insight: illegal activities within organisations can often remain concealed for an extended period not because those involved isolate themselves from the world, but rather because these actors and their illicit activities are socially embedded. This social embedding occurs when individuals engage in wrongdoing and are part of social groups and networks that provide cover and protection for their activities.

In other words, secrets and illegal activities persist because they are shielded by the social connections and networks that individuals maintain. These networks can help conceal wrongdoing by providing a shield of trust and cooperation that makes it difficult for external parties to uncover or expose these activities. This concept underscores the importance of understanding the social dynamics and relationships that contribute to keeping illicit activities hidden.

The process of "making things work" on megaprojects is not dissimilar to the challenge of crime control. On megaprojects, team members often employ short-term tactics that, over time, coalesce into strategies. These strategies are emergent constellations of tactics that are strategically bundled together based on the available resources. This "muddling through" process is significant to understand because these practical ways of thinking and acting become the form, mundane, routine, and largely hidden aspects of the megaproject decision-making environment.

As DeLillo (1977) highlighted, "everyday things represent the most overlooked knowledge". Governmentality theory serves as a valuable conceptual tool to uncover and reveal the taken-for-granted and otherwise hidden practices of everyday decision-making in megaprojects.

Chapter summary

This chapter presented an alternative perspective for understanding power dynamics within megaproject decision-making. Governmentality theory was highlighted as a relevant approach for examining how both megaproject clients and team members engage in governing and are governed. This analysis focuses on the rationalities, tactics, strategies, choices, desires, and aspirations that shape the leadership, management, and direction of individuals and organisations involved in megaprojects.

In this perspective, the exercise of power through various deviant tactics is not inherently negative, contrary to traditional notions of power. Instead, this book acknowledges that the practice of deviance can lead to both positive and negative consequences within the context of megaprojects. By adopting this framework, a more nuanced and comprehensive understanding of power relations in megaproject decision-making can be achieved.

References

Biesenthal, C., Clegg, S.D., Mahalingam, A., & Sankaran, S. (2018). Applying institutional theories to managing megaprojects. *International Journal of Project Management*, 36(1), 43–54.

Bodankin, M., & Tziner, A. (2009). Constructive deviance, destructive deviance and personality: How do they interrelate? *Amfiteatru Economic*, 11(26), 549–564.

Boyce, G., & Davids, C. (2004). The dimensions of governmentality studies in accounting: Complementary and critical potentials. In *Fourth Asia Pacific Interdisciplinary Research in Accounting Conference*, Singapore, July, APIRA and Singapore Management University.

Brislin, R. (1991). *The art of getting things done: A practical guide to the use of power*. New York: Praeger Publishers.

Carroll, A., & Buchholtz, A. (2006). *Business & society: Ethics and stakeholder management*. Mason, OH: Thomson/South-Western.

Clegg, S.R. (1975). *Power, rule and domination*. London: Routledge and Kegan Paul.

Clegg, S.R. (1989). *Frameworks of power*. London: Sage.

Clegg, S.R. (2014). Circuits of power/knowledge. *Journal of Political Power*, 7(3), 383–392.

Clegg, S.R., Pitsis, T.S., Rura-Polley, T., & Marosszeky, M. (2002). Governmentality matters: Designing an alliance culture of inter-organizational collaboration for managing projects. *Organization Studies*, 23, 317–337.

Chan, P., & Jia, A. (2019). Finding the micro in the mega: A multi-ethnographer study into the microdynamics in a mega construction project. In C. Gorse, & C.J. Neilson (Eds.), *Proceeding of the 35th Annual ARCOM Conference*, 2–4 September 2019, Leeds, Association of Researchers in Construction Management, 1–10.

Dean, M. (2010). *Governmentality: Power and rule in modern society*. London: Sage Pubs.

DeBlois, M., Herazo-Cueto, B., Latunova, I., & Lizzeralde, G. (2011). Relationships between construction clients and participants of the building industry: Structures and mechanisms of coordination and communication. *Architectural Engineering and Design Management*, 6(1), 1–20.

de Certeau, M. (1984). *The practice of everyday life*. Berkeley: University of California Press.

DeLillo, D. (1977). *Players*. New York: Alfred Knopf.

Donzelot, J. (1979). *The policing of families*. New York: Pantheon Books.

Ewick, P., & Silbey, S. (1998). *The common place of law: Stories from everyday life*. Chicago, IL: The University of Chicago Press.

Ewick, P., & Silbey, S. (2003). Narrating social structure: Stories of resistance to legal authority. *American Journal of Sociology*, 108(6), 1328–1372.

Fiori, C., & Kovaka, M. (2005). *Defining megaprojects: Learning from construction at the edge of experience*. ASCE Construction Research Congress: Broadening perspectives, April 5–7, California.

Fiske, S., & Depret, E. (1996). Control, interdependence and power: Understanding social cognition in its social context. *European Review of Social Psychology*, 7(1), 31–61.

Flyvbjerg, B. (1998). *Rationality and power: Democracy in practice*. Chicago, IL: The University of Chicago Press.

Foucault, M. (1979). Governmentality. *Ideology and Consciousness*, 6, 5–21.

Foucault, M. (1980). Body/power. In C. Gordon (Ed.), *Power/knowledge: Selected interviews and other writings, 1972–1977*. London: Penguin.

Foucault, M. (1991). Questions of method. In G. Burchell, C. Gordon, & P. Miller (Eds.), *The Foucault effect: Studies in governmentality* (pp. 73–86). Chicago, IL: Chicago University Press.

French, J., & Raven, B. (1959). The bases of social power. In D. Cartwright (Ed.), *Studies in social power* (pp. 150–167). Ann Arbor: University of Michigan.

Galbraith, J. (1983). *The anatomy of power*. London: Corgi Books.

Garland, D. (1997). Governmentality and the problem of crime: Foucault, criminology and sociology. *Theoretical Criminology*, 1(2), 173–214.

Goffman, E. (1961). *Asylums*. New York: Anchor Books.

Gordon, C. (1991). Governmental rationality: An introduction. In G. Burchell, C. Gordon, & P. Miller (Eds.), *The Foucault effect: Studies in governmentality* (pp. 1–52). Chicago, IL: University of Chicago Press.
Hallgren, M., & Soderholm, A. (2010). Orchestrating deviations in global projects: Projects-as-practice observations. *Scandinavian Journal of Management*, 26, 352–367.
Hajikazemi, S, Aaltonen, K., Ahola, T., Aarseth, W., & Andersen, B. (2020). Normalising deviance in construction project organizations: A case study on the collapse of Carillion. *Construction Management and Economics*, 38(12), 1122–1138.
Haugaard, M. (2012). Editorial: Reflections upon power over, power to, power with, and the four dimensions of power. *Journal of Political Power*, 5(3), 353–358.
Karlberg, M. (2004). *Beyond the culture of contest: From adversarialism to mutualism in an age of interdependence.* Welwyn: George Ronald Publisher.
Lemke, T. (2000). Foucault, governmentality and critique. In *Rethinking Marxism Conference, University of Amherst*, September 21–24.
Loosemore, M. (1998). The influence of communication structure upon management efficiency. *Construction Management and Economics*, 16(6), 661–671.
Marshall, N. (2006). Understanding power in project settings. In D. Hodgson, & S. Cicmil (Eds.), *Making projects critical* (pp. 207–231). London: Red Globe Press.
Morris, P., & Hough, G. (1987). *The anatomy of major projects*. London: Major Projects Association.
Munro, L. (2000). Non-disciplinary power and the network society. *Organisation*, 7(4), 679–695.
Ninan, J., Mahalingam, A., & Clegg, S. (2019). External stakeholder management strategies and resources in megaprojects: An organizational power perspective. *Project Management Journal*, 50(6), 625–640.
Pinto, J.K. (1996). *Power and politics in project management.* Newtown Square, PA: Project Management Institute.
Pinto, J.K. (2014). Project management, governance, and the normalization of deviance. *International Journal of Project Management*, 32(2014), 376–387.
Robinson, S., & Bennett, R. (1995). A typology of deviant workplace behaviors: A multidimensional scaling study. *The Academy of Management Journal*, 38(2), 555–572.
Rose, N., & Miller, P. (1992). Political power beyond the state: Problematics of government. *British Journal of Sociology*, 43, 173–205.
Rose, N., O'Malley, P., & Valverde, M. (2009). Governmentality. *Legal Studies Research Paper No. 09/94*, The University of Sydney.
Sayer, A. (2001). For a critical cultural political economy. *Antipode*, 33(4), 687–708.
Scott, J. (1985). *Weapons of the weak.* New Haven, CT: Yale University Press.
Sending, O., & Neumann, I. (2006). Governance to governmentality: Analyzing NGOs, states, and power. *International Studies Quarterly*, 50, 651–672.
Siva, J., & Gajendran, T. (2023). The hidden barriers to social value delivery: Investigating the decision-making environment. *Built Environment Project and Asset Management*, 13(4), 610–628.
Turner, J. (2005). Explaining the nature of power: A three-process theory. *European Journal of Social Psychology*, 35, 1–22.
Van de Bunt, H. (2010). Walls of secrecy and silence: The Madoff case and cartels in the construction industry. *Criminology and Public Policy*, 9(3), 435–453.
van Marrewijk, A. (Ed.) (2015). *Inside megaprojects: Understanding cultural practices in project management*. Copenhagen: CBS Press.

van Marrewijk, A., Clegg, S.R., Pitsis, T., & Veenswijk, M. (2008). Managing public-private megaprojects: Paradoxes, complexity and project design. *International Journal of Project Management*, 26, 591–600.

Vaughan, D.S. (1996). *The challenger launch decision: Risky technology, culture, and deviance at NASA*. Chicago, IL: University of Chicago Press.

Warner, J. (2010). Constructing negotiations: bargaining, learning and fighting. In E. Chinyio, & P. Olomolaiye (Eds.), *Construction stakeholder management* (pp. 138–158), Oxford: Wiley-Blackwell.

Warren, D. (2003). Constructive and destructive deviance in organizations. *The Academy of Management Review*, 28(4), 622–632.

Weber, M. (1947). *The theory of social and economic organization*. New York: Free Press.

Wild, A. (2002). The unmanageability of construction and the theoretical psycho-social dynamics of projects. *Engineering Construction and Architectural Management*, 9(4), 345–351.

Winch, G. (2017). Megaproject stakeholder management. In B. Flyvbjerg (Ed.), *Oxford handbook of megaproject management* (pp. 339–361). Oxford: Oxford University Press.

Wrong, D. (1979). *Power: Its forms, bases and uses*. New York: Harper & Row.

4 Governmentality framework for decision-making

The governmentality framework for megaproject decision-making

The interweaving of governing mechanisms, deviance tactics, and power relations makes it challenging to define an optimal environment for megaproject decision-making. In other words, a "one size fits all" approach does not work when managing and delivering megaprojects. The governmentality framework for decision-making is not about offering a guaranteed recipe for success. Instead, it provides a framework that can enhance the likelihood of success in a variety of decision-making settings.

First, megaproject decision-making is distinguished by its alignment with a governmentality perspective. This approach encompasses a diverse blend of governance mechanisms and an array of governing practices. To effectively navigate the complexity and uncertainty inherent in megaproject decision-making, a blended diagram of power is essential. This necessitates, first, recognising and accepting the coexistence of formalised governance mechanisms alongside informal governing practices. Second, it requires designing megaproject governance frameworks that accommodate both these rationalities. Embracing a governmentality approach to decision-making carries significant implications for megaproject management. It particularly influences our understanding of how the practice of deviance might be more effectively managed and directed towards enhancing project performance.

Second, the concept of deviance in megaprojects is closely linked to the exercise of power through various tactics, which can be broadly classified as either constructive or destructive. Despite the rigid governance structures that typically characterise megaprojects, factors such as energy, experience, skill, resources, and strategy play a crucial role in shaping decision-making practices. Within the confines of established rules, there is significant scope for action. The inherent complexities of such projects invariably introduce elements of uncertainty into the decision-making process. Team members, akin to experienced strategists in a game, adeptly deploy a range of deviant tactics to achieve diverse objectives.

However, it is important to note that the exercise of power through these tactics is not inherently negative, as is often the case in traditional perceptions of power dynamics. The practice of deviance in megaprojects can lead to both positive and negative outcomes. More significantly, actions that might have traditionally been

DOI: 10.1201/9781003377481-4

viewed as destructively deviant, could in the context of megaprojects, prove to be constructive, and vice versa. This nuanced understanding of deviance highlights the complexity in megaproject decision-making and the need for a more balanced approach to managing such practices.

To illustrate these components of the governmentality framework, we will focus on two main megaprojects in Asia: a Sports Facility megaproject and an Airport megaproject.

The Sports Facility megaproject

The Sports Facility megaproject is approximately US$1 billion fully integrated multi-use, multi-sport, and entertainment hub in Asia. The Sports Facility megaproject is located on a 35-hectare site. It is a megaproject built with the community in mind. It aimed to encourage large numbers of people from all demographics – from elite to novice – to adopt and pursue sports, making sports part of their lives. It was also conceived to draw international events to its world-class facilities comprising multi-purpose venues with a variety of seating capacities, which was part of the government's broader vision of creating a thriving sports, entertainment, and lifestyle ecosystem.

The Sports Facility was officially declared open in 2015. It was the largest sports facility infrastructure public–private partnership (PPP) project in the world2. The project was also the first and largest PPP project in the country. The multi-award-winning Sports Facility is the first of its kind to combine world-class infrastructure for various sports activities with community lifestyle and entertainment facilities.

The project involved the demolition of an existing National Stadium and the construction of new buildings, including a national stadium, a multi-purpose indoor arena, an aquatic centre, a water sports centre, an information and resource centre, a sports promenade, and a commercial space/retail mall. It also incorporated an existing multi-purpose indoor arena, an Indoor Stadium, which was built in the late 1980s. The centrepiece of the new Sports Facility is the 55,000-seat National Stadium, which features the world's largest free-spanning dome, a retractable roof, and an innovative low-energy bowl cooling system.

A special purpose vehicle (SPV) was formed to finance, design, build, and operate the facility in partnership with the government agency client over a 25-year period. The SPV invested in equity and utilised bank debt financing to build the facilities. Upon completion of the contract, the facility was handed over to the government agency.

The government agency was formed in 1973 and is the lead agency tasked with developing a holistic sports culture for the country. The agency is under the authority of the Ministry of Community Development, Youth, and Sports and is made up of seven groups (see Figure 4.1). There are approximately 800 employees in the organisation. The Sports Infrastructure unit within the Sports Facilities Group is the unit responsible for procurement of projects such as the Sports Facility megaproject.

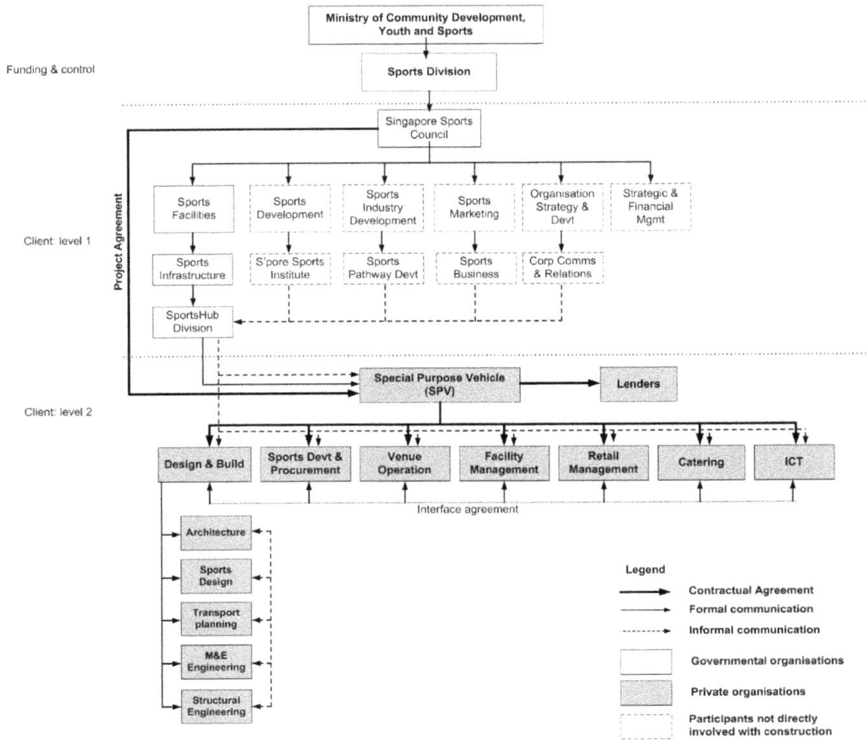

Figure 4.1 The public–private partnership project organisational structure of the Sports Facility megaproject (Source: Siva & London, 2016).

Box 4.1 Case study – formal PPP structure of the Sports Facility megaproject.

In the case of PPP projects, it is common to see a network of multiple clients, as demonstrated in the Sports Facility megaproject. This particular project involved a diverse group of clients: the government agency, which was tasked with establishing the project framework, and various partners in the PPP consortium, each playing a unique role in the project's development (refer to Figure 4.1 for details).

All the partners were united under the terms of the Project Agreement and adhered to the protocols outlined in the Interface Agreement. The Chief Operations Officer of the SPV managing the Sports Facility megaproject remarked on the exceptional detail and structure of the project's contract documentation, noting it as one of the most thorough he had encountered.

Figure 4.1 illustrates the intricate PPP structure of the Sports Facility megaproject. This diagram details the key stakeholders, their roles, and the relationships between them. The government agency and the SPV, responsible for project coordination, established formal structures and protocols. These frameworks were designed to ensure accountable decision-making while respecting the roles and boundaries of the involved parties.

Given the megaproject's complexity, a dedicated division was created solely for this project. The Sports Facilities Division, consisting of 15 employees, focused on overseeing contract compliance and facilitating coordination between the broader government agency and SPV. This division leverages the expertise from other groups within the government agency, like the Sports Development Group, to define the project's scope accurately. At the helm of the Sports Facility Division was a Project Director, who managed three sub-divisions: operations and programming, design, and technical and contract compliance.

Box 4.2 Case study – formal governance mechanisms for the Sports Facility megaproject.

The Sports Facility Division played a crucial role in the project's success, as it was heavily relied upon for thorough supervision and work surveillance of the project team. This was essential to ensure strict adherence to the contract's terms. The Interface Agreement, comprising a complex network of procedures, protocols, and processes, functioned as a tool for the client to exercise significant control over other stakeholders and project team members. This agreement was particularly important for maintaining transparency and accountability. The effectiveness of these protocols and procedures was demonstrated by their ability to provide clear traceability. They established well-defined lines for the exchange of official documents between members of the project team. The following quotes illustrate this aspect further:

> Yes because they'll [the consultants] say that, "oh we find that this is good". But we say "even though its good…the project agreement requires such" and it requires them to provide certain details because it's captured in their reports and their reports is basically how we track their performance.
>
> (Deputy Director, Operations and Programming, Sports Facilities Division, Sports Facility megaproject)

> We have to think about what is written in the contract and what does the contract say, you know...So it's the contract that manages the job.
>
> (Chief Operations Officer, Special Purpose Vehicle, Sports Facility megaproject)

The control and surveillance mechanisms employed in the Sports Facility megaproject are reminiscent of the disciplinary techniques discussed in Chapter 3, particularly those associated with sovereign power and the Panopticon model. These concepts of power and control provide a useful context for understanding the structured approach used in managing the Sports Facility megaproject.

The Airport megaproject

Case study 2 examines a dedicated low-cost carrier terminal located in Asia with an estimated budget of US$0.9 billion. The Airport megaproject stands out as the largest purpose-built mega terminal specifically designed to cater for low-cost carriers. Its construction aimed to sustain the remarkable growth of the country's low-cost travel sector, which initially gained momentum with the launch of the first low-cost carrier terminal. The primary objective of this project was to establish the nation's next-generation international airport hub, facilitating seamless connectivity between local and international low-cost and full-service carriers.

The Airport megaproject originated as part of the country's National Airport Master Plan in 2008. It was jointly undertaken by a consultant involved with a previous airport project and an airport consultancy and engineering firm based in The Hague, Netherlands. The funding for this national infrastructure project was entirely provided by the airport operator through *Sukuk*, which are Islamic bonds. These Sukuk were subscribed to by a diverse group of investors, including government agencies, corporate entities, and financial institutions. The airport operator is responsible for overseeing the operations, management, development, and maintenance of 39 airports across the country, encompassing both aeronautical and non-aeronautical components.

This project was a crucial component of the country's Economic Stimulus Package implemented during the global financial crisis in 2009. It engaged the expertise of over 50 local consultant and contractor firms across 40 different work packages. The coordination of such a large number of packages and consultant firms presented a considerable logistical challenge. However, it was a strategic decision that aligned with the government's goal of providing support to as many local firms as possible during the economic downturn. This approach not only helped stimulate the local economy but also allowed these firms to gain valuable experience by participating in a world-class megaproject.

The Airport megaproject was executed as a design and build project utilising an accelerated or "fast track" approach. This approach was adopted in response to the escalating congestion issues experienced at the previous low-cost carrier terminal and to meet the growing demands of the main airline. The selection process for the architectural firm responsible for designing the main terminal building, a pivotal component of the airport, involved an international competition held in 2009. Initially, 12 entries were submitted, which were subsequently narrowed down to a final shortlist of five.

The Airport megaproject encompasses several key components, including the main terminal building, a skybridge, a satellite building, interconnected piers, and an integrated shopping mall with an attached car park facility. Notably, over 55,000 square metres of space within the terminal building is allocated for commercial enterprises.

The Airport megaproject was purpose-built to cater to the needs of 45 million passengers annually. Moreover, its design incorporates provisions for future expansion to align with evolving trends in air travel and the operational requirements of airlines.

The construction of the Airport megaproject garnered significant media attention due to a series of controversial delays and cost overruns. The opening of the Airport megaproject faced no less than five delays beyond its initial target completion date of September 2011. Ultimately, in May 2014, the Airport megaproject was officially inaugurated, serving as a replacement for the previous low-cost carrier terminal.

Box 4.3 Case study – the evolving project management of the Airport megaproject.

In 2009, the Ministry of Finance directed the airport operator to engage XXX Consultancy Services as the project management consultant. The consultancy was tasked with managing various aspects of the project, including design, construction, consultancy, and the tendering processes. To ensure effective project management, a dedicated team comprising approximately 100 staff members was established within the consultancy. This team played a pivotal role in the day-to-day oversight of the project. The consultancy also assumed the overarching project management responsibilities, overseeing the performance of all consultants and coordinating the various project packages as needed.

Furthermore, within the client organisation, a separate team known as the airport client office was established to oversee the project management team. Comprising approximately 20 personnel, this team played a supportive role, with a specific focus on advancing payment claims submitted by the project management consultant firm to the client office. Both the consultancy and airport client office teams were stationed at the existing operating international airport. A temporary site office was set up within the airport's car park area, where weekly progress meetings were conducted to enable the project management consultant to provide regular updates to the client regarding various project-related matters.

In 2011, when the consultancy's contract expired, the airport operator made the decision not to extend their services as the consultant firm responsible for overseeing the project management of the Airport megaproject. The Chief Operating Officer of the airport operator cited dissatisfaction with their performance as the main reason for this decision. Consequently, the airport client office assumed full project management responsibilities. To bolster the airport client office team, the airport operator increased the team's size to over 80 members by recruiting individuals from both within the airport operator and externally. Notably, some team members were brought in from the previous project management consultancy firm. This move was aimed at ensuring continuation of project management.

Box 4.4 Case study – formalised structures and protocols used in the Airport megaproject.

The airport operator client implemented well-defined formalised structures and protocols to facilitate project coordination for the Airport megaproject. These protocols delineated the roles and responsibilities of different parties involved in the project and established official channels for document exchange among project stakeholders (see Figure 4.2).

The airport client office played a crucial role within the client organisation by identifying the technical and operational requirements of various divisions within the airport operator. The Client Design Consultant (CDC), led by the architecture firm responsible for Package 1, served as the client's representative and supervision team to ensure that the contractor's work aligned with the client's requirements. Essentially, the CDC can be viewed as a client on a different level, as it monitors and supervises the contractor's activities. A similar structure was established for each of the 40 work packages, although this case study primarily focuses on the first work package, which was the largest and most significant. In each package, a CDC team was responsible for overseeing the work of other consultants and contractors on behalf of the client.

Project team members characterised their decision-making environment as highly regulated, with project-related matters strictly governed by formally prescribed mechanisms. These mechanisms included progress reports, project review meetings, and project approval procedures and processes. They frequently mentioned the intricate network of "rules", "meetings", "processes", and "paperwork" that clearly outlined their roles and provided guidelines for dealing with day-to-day matters on the project.

> So how this matrix works… I have a specific person who shall I interact with…I will go to my superior and similarly my down liner, they

understand their level. They are not supposed to just do anything… So this is how we work…what meetings I go to or don't go, it's all defined.

(Senior Construction Manager, Airport Client Office, Airport megaproject)

There are a lot of meetings…Meetings and paperwork…we have to go through all the process[es].

(Senior Project architect, Client Design Consultant – Architecture firm, Airport megaproject)

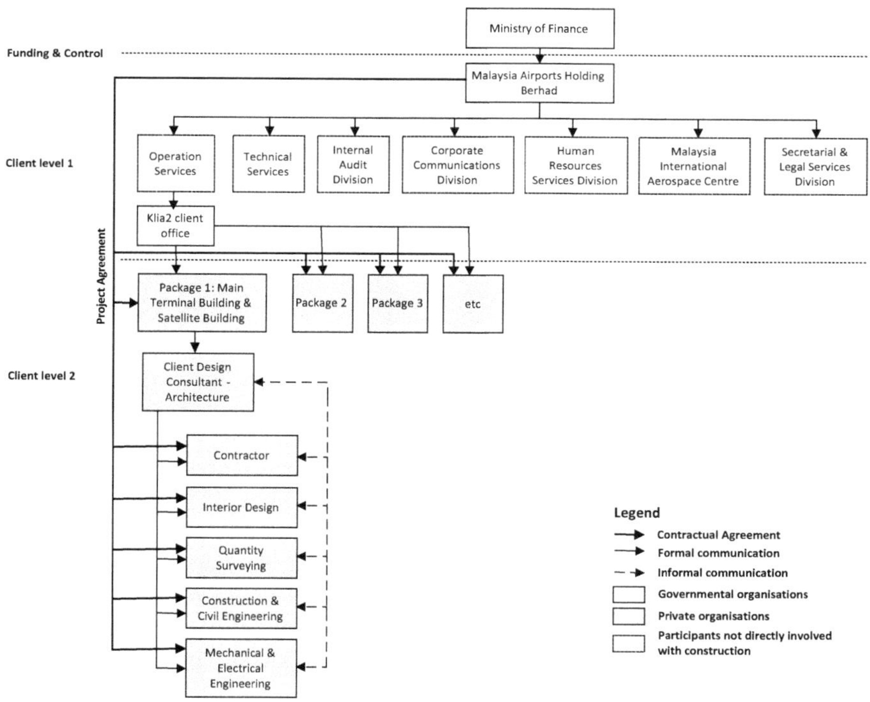

Figure 4.2 Formal project organisational structure for the Airport megaproject.

A balance of governance and governing

Megaproject governance involves a blend of formal mechanisms and informal practices, including various tactics of deviance. Formally prescribed governance, embodied in project contracts, policies, and procedures, sets expected standards for behaviour. Conversely, team members also navigate governance at an instrumental level, engaging with a network of actors, organisations, and norms. This dual approach shapes how they manage daily unplanned and spontaneous situations, balancing structured protocols with more flexible, adaptive practices to address the practical needs of project execution.

There is an inherent tension between the formal project frameworks and the practical realities of how decision-making is undertaken on megaprojects. The pluralistic nature of megaproject governance and the uneven distribution of power often pose challenges to the implementation of standard project management routines in decision-making (Ninan et al., 2019; Biesenthal et al., 2018).

Megaprojects are marked by complex and intertwined power dynamics, political processes, and conflicting interests (Marshall, 2006; Pinto, 1996). This complexity renders decision-making processes in such projects too intricate to be effectively governed solely by formal mechanisms based on disciplinary power. Box 4.5 presents insights into the different forms of authority and power that exist within the context of megaprojects. It delves into the dynamics between the formal project organisational structures and the practical implementation of governance policies.

Box 4.5 Blending governance and governing mechanisms in megaproject decision-making.

Even with the presence of sophisticated information technology systems, project team members emphasise the importance of physical proximity in megaproject decision-making. Being close enough to hold regular face-to-face meetings is seen as invaluable. It allows for spontaneous gathering, where team members can "solicit agreements" and "make sure they're all on the same page". These face-to-face interactions provided a platform for informal exchanges. Project team members found that being able to quickly share ideas, gather opinions, and gain consensus in an informal setting was instrumental in shaping decisions before they were formally agreed upon.

> There's a lot of running downstairs you know... So face-to-face and then [formalise] and that's why they're in the same building as we are.
>
> (Senior Design Manager, Design and Build Contractor, Sports Facility megaproject)

> In terms of interfacing cause there's a lot of nuances in terms of where its body language or the way you speak...all these sorts of things I find it depends on who I'm speaking to it just changes.
>
> (Senior Project and Design Manager, Architecture firm, Sports Facility megaproject)

For these project team members, informal communication and decision-making often occurred outside the predefined structures and processes.

This is evidenced by their remarks, such as "there's a lot of running downstairs" and "its body language or the way you speak". These comments highlight the frequency and importance of informal interactions in their daily work.

While the formalised methods of communication and established protocols for document exchange are undoubtedly important, project team members pointed out the need for a balanced approach. Combining formal and informal communication methods was seen as essential, especially when it came to working through project issues and "soliciting agreements". This balance allows for more efficient and effective decision-making, blending the structure of formal processes with the agility and personal touch of informal communication.

As demonstrated in the case studies, project team members typically demonstrate respect for rules, technical compliance, and hierarchical authority. However, they also recognise that strict adherence to rigid protocols often hampers effective problem-solving. Therefore, deviations from these protocols, though seen as departures from the ideal, are deemed necessary for pragmatic decision-making and progress in complex project environments. This duality illustrates the nuanced balance between following established procedures and adapting flexibly to the realities of megaproject decision-making.

Box 4.6 The strategic interplay of rules and flexibility to navigate governance.

While a traditional view of governance views policies and guidelines as constraints that limit or even prevent certain actions, an alternative perspective treats these rules more like a game. In this view, rules don't just restrict; they create opportunities and "loopholes" where deviations and strategic advantages can emerge. Project team members in the case studies frequently encountered situations where strictly following the rules wasn't the most effective approach. In these cases, they recognised the necessity to adapt, often resorting to deviant strategies to reshape outcomes.

The stories shared by project team members from the case studies reveal a complex interplay with the established rules of governance. They spoke of navigating these systems using a variety of tactics as indicated by their use of terms like "compensate", "convince", "trust", "loopholes", "argue", "grey area", "smooth things over", "understanding concerns", "game", "pretend", "ignore", "backdoor it", and "behind the scenes". These terms reflect their experiences and strategies in dealing with governance mechanisms, highlighting the flexibility and creativity often required in these processes.

Decision-making in megaprojects often involves a complex mix of power structures and governing principles. This field is characterised by a diverse array of diagrams and ontologies of power, each bringing its own set of governing mechanisms and practices to the table. The convergence of these different approaches creates a detailed and intricate network of policies, practices, and processes. This network is so multifaceted that it cannot be distilled into a simple, one-size-fits-all formula.

In Chapter 3, we delved into a discussion of the three distinct models of power: sovereign, Panopticon, and network. The two megaprojects shed light on the various strategies employed by clients and project team members in megaprojects. These strategies demonstrate similarities to both the disciplinary and non-disciplinary approaches found in all three of these power models. The megaprojects reveal how these diverse tactics were used to adapt to, create, harness, and cultivate power within the complex dynamics of megaprojects.

The decision-making environment in megaprojects significantly resembles the Panopticon and network diagrams of power, as outlined in Table 3.2 (see Chapter 3). On the one hand, there are disciplinary mechanisms akin to the sovereign and Panopticon diagrams, whereby a central authority wields power. On the other hand, governing authorities comprise not only those in formalised structures but also those engaged in less visible, informal negotiations, and dealings. This latter aspect aligns with the characteristics of the network diagram, emphasising the importance of both visible and hidden interactions in problem-solving and decision-making within megaprojects.

Box 4.7 Lesson – a governmentality framework of power is required for megaproject decision-making.

In the decision-making processes of the two megaproject case studies, elements of both Panoptic and sovereign power were observed within the broader framework of network power dynamics. This mix of power relations highlights how formally prescribed governance mechanisms were not the only guiding forces. Alongside these established protocols, a variety of less formal governing practices were also in play. These included an array of tactics that could be categorised as deviant.

The case studies revealed a nuanced diagram of power within megaprojects, blending both disciplinary and non-disciplinary mechanisms. This amalgamation forms what can be described as a governmentality framework. It is a complex structure that encompasses a spectrum of behaviours, ranging from complete compliance to various degrees of deviation from the norm. Figure 4.3 illustrates a simplified version of this governmentality diagram. In this model, certain participants are isolated, subject to constant surveillance, while others are interconnected and able to communicate freely through different channels.

This multi-layered power structure is characterised by some nodes being isolated and strictly controlled, but with the potential for any node to connect with others in the network. As depicted in Figure 4.3, despite the presence of a formal central authority (indicated by the largest node) exerting legitimate power over others (shown with solid lines), a network of informal relationships (indicated by dashed lines) can still form between various actors. This indicates the coexistence of formal, centralised control and informal, decentralised interactions within the megaproject's governmentality framework.

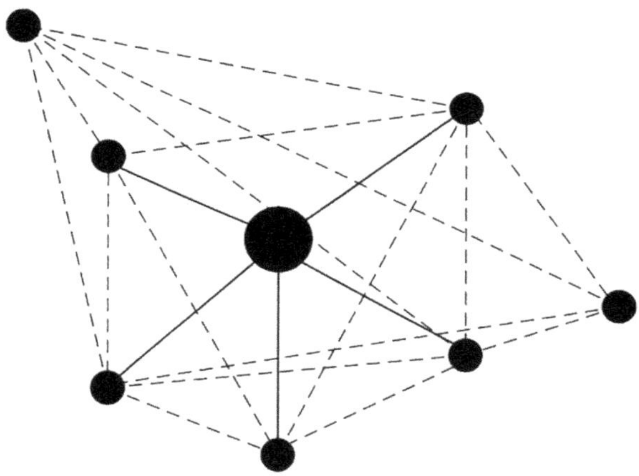

Figure 4.3 A governmentality framework of power in megaproject decision-making.

Omnipresence of power

Examining the case studies through a governmentality lens reveals that various authorities exert influence in different areas, often in response to competing interests. Traditional views of power suggest a linear structure, with power centralised in a single, dominant entity. However, the blended diagram of power (depicted in Figure 4.3) illustrates a more nuanced reality for the two megaprojects under study. Here, governing processes and practices are seen as dynamic and dispersed, rather than only centralised.

This concept aligns with Foucault's idea of the omnipresence of power, suggesting that power exists in diverse locations, or in other words, is "everywhere". The stories shared by project team members emphasise this point. They reveal that positional power, which is more static and centralised, coexists with a dynamic, relational form of power. This relational power is process-based and manifests in the decision-making practices of the megaprojects, illustrating a more fluid and distributed approach to governance.

Box 4.8 The active roles and influence of team members in shaping project outcomes.

Through the lens of governmentality, an interesting pattern emerged in the two case studies: different actors took on central roles at various stages of the projects. Certain project team members were instrumental not just in guiding their peers towards compromises and agreements but also in crafting the decision-making process itself. They often employed deviant tactics to effectively influence and challenge the status quo.

In the Sports Facility megaproject, the client played a pivotal role in achieving the approval of an innovative bowl cooling system. They achieved this by bypassing several layers of bureaucratic authority, leveraging their positional power to move the process forward. Meanwhile, in the Airport megaproject, the Senior Design Coordinator from the design and construct contractor firm emerged as a key influencer in decision outcomes. His adeptness at navigating the decision-making environment was evident in the stories he shared. He demonstrated proficiency in using various tactics to address different challenges – from identifying contractual loopholes and feigning ignorance to disrupting hierarchical norms. These strategies enabled him to turn situations to his advantage, highlighting the flexibility and adaptability required in such roles.

Far from being mere passive recipients of rules and regulations, various members of the megaproject teams – including the client, design team, and contractor – often took active roles in the decision-making process. On numerous occasions, these individuals and groups demonstrated their ability to exercise rights and autonomy. They skilfully navigated complex situations to address the diverse and sometimes conflicting interests of both formal and informal stakeholders.

This perspective portrays megaproject team members as both subjects to regulatory frameworks and as independent agents with the ability to make autonomous decisions. This dual role challenges the traditional view of these individuals as simply being regulated by external forces. Previous research on megaprojects has often overlooked the significant influence that these varied "governmental authorities" and "everyday agents" exert in decision-making and in shaping the outcomes of megaprojects.

The practice of deviance in megaproject decision-making

Members of megaproject teams are acutely aware of the complexities and uncertainties inherent in such complex projects, which significantly impact their decision-making. They navigate these challenges much like skilled strategists in a game, employing a range of tactics to meet various objectives. These tactics, shaped by diverse and sometimes conflicting interests, are integral to decision-making in megaprojects.

In this context, decision-making in megaprojects can be likened to a strategic game, bound by rules and constraints, where outcomes are heavily influenced by the players' skills, experience, and resources. This analogy emphasises the importance of a governmentality framework for megaproject decision-making, which requires a balanced approach that considers both governance structures and the practical aspects of governing.

While megaprojects present a wide array of challenges and issues that require resolution, the responses of clients and project team members to these challenges are not boundlessly diverse. Instead, team members typically draw from a shared set of tactics and power strategies. These approaches are informed by historical responses and experiences, shaping a collective understanding of problem-solving in this context. This shared repertoire can be broadly categorised into two types of deviance: constructive and destructive.

Consequently, the choices available to project team members are not unlimited. They are confined to the strategies and solutions that have been previously employed and understood within their professional community. This means that individual team members do not create entirely unique approaches to the decision-making environment; rather, their actions are influenced by established norms and practices.

The concept of deviance in the context of project management is rooted in the idea that all transactions operate under certain rules or systems. These rules and systems inherently create opportunities for both adherence and resistance (Ewick & Silbey, 2003). Beckert (1999) further suggests that any ambiguities or gaps in these systems can be exploited by actors for their own benefit. This manipulation of rules can lead to different outcomes, depending on how it is applied.

When deviance is harnessed constructively, it can facilitate positive outcomes for the project, leveraging the flexibility within the system to overcome challenges or innovate. However, when deviance is applied destructively, it can lead to negative consequences, potentially derailing project objectives. Table 4.1 provides

Table 4.1 Categories of constructive and destructive deviance on megaprojects

	Constructive	*Destructive*
Deviation	Departs from prescribed standards in terms of expected behaviour	Departs from prescribed standards in terms of expected behaviour
Motivation	Collective interests for the benefit of the project; unified motivation	Personal interests for individual or organisational benefit, divided/ conflictual motivation
Outcome	Beneficial to the megaproject, i.e. positive outcomes such as improved performance, resolution of issues, time/cost savings	Harmful to the megaproject, i.e. negative outcomes such as delays, increased costs, reduced performance, or objectives not achieved
Characteristics	Collaborative, altruistic, ethical, leadership	Opportunistic, self-centred, unethical, deceptive, manipulative

a concise summary of the defining features of these two forms of deviance in megaprojects – constructive and destructive.

Constructive deviance

Constructive deviance in megaprojects involves intentional behaviours that, while deviating from prescribed standards, ultimately benefit the project. This form of deviance employs tactics aimed at achieving positive outcomes, such as disrupting established hierarchies, negotiating swaps, leveraging friendship or social ties, and gatekeeping. Project team members shared instances where formal procedures and structures fell short in effectively resolving intricate project disputes and issues. In these situations, they often turned to alternative solutions, which included bending or circumventing the established rules and hierarchies with the project's organisational framework.

Box 4.9 Case study – deviating from rules to benefit the project.

Members of project teams recounted instances where they intentionally bypassed established protocols to enhance project performance. Despite the existence of well-defined project protocols and procedures, they highlighted a notable discrepancy between the formal framework and the actual practices observed, particularly in terms of communication and problem-solving. This gap became especially evident when addressing emerging project issues that required prompt and pragmatic resolutions.

A key factor noted was the rapid pace of decision-making required in these megaprojects. The urgency and speed with which decisions needed to be made often led to a compressed planning horizon. As a result, team members frequently found themselves taking spontaneous actions that deviated from the formal protocols.

The Chief Operations Officer of the airport operator in the Airport megaproject emphasised the importance of swiftly resolving issues in the context of a fast-paced project, stating, "the project is moving fast and a decision is required…because if you don't make a decision there will be idling costs and people are waiting". This urgency often led team members to prioritise informal channels for problem resolution, resorting to formal procedures only when absolutely necessary.

Team members in the case studies also demonstrated how deviating from standard protocols could lead to improved project outcomes, particularly through the tactic of "gatekeeping" in managing information flow. This approach involved selectively withholding or filtering information as a strategic measure. They emphasised the criticality of providing the "right information" at the "right time" and in the "right manner". This strategy was not just about controlling information access but also about strategically managing its dissemination to influence project outcomes positively.

The behaviours of project team members were primarily driven by a focus on outcomes and a practical need to keep the project on track. As a result, they sometimes adopted strategies that led to the partial implementation of policies and procedures. Their primary goal was to ensure the smooth progression of the project, even if it meant bending the established rules.

Project team members often exhibited collaborative and leadership qualities in their application of constructive deviance. This approach was fundamentally motivated by a commitment to positively impact the project.

Box 4.10 Case study – leadership to champion causes.

The leadership abilities of certain individuals played a pivotal role in fostering collaboration and advancing project goals in both case study megaprojects. Project team members shared experiences showing their proactive engagement with influential individuals in positions of power to advance the project. These key figures were acknowledged for their ability to sway opinions and guide the project. Brislin (1991) referred to this as leveraging a "knowledge base". In this context, project team members relied on their understanding of who within the project had the necessary expertise or authority to support their initiatives. This strategic use of knowledge about influential individuals was instrumental in leading teams towards compromises.

> I think that when you work on a big project there's a lot of people around the interpersonal relationship and the people who provide a bit of leadership to glue that all together are really important because it is a long journey and people do come and go and people do also rub off people in the wrong way at times.
>
> (Senior Associate, Sports Architecture firm, Sports Facility megaproject)

> There's a little bit of understanding who's who, who's the right person to talk to at the right time and picking the right time to be letting them know what's going on and if there's something they can help us to getting something changed.
>
> (Senior Project and Design Manager, Architecture firm, Sports Facility megaproject)

> Sometimes we have to use other people to help us get the message across and that's where someone like Ben (Senior design manager) has

a great degree of influence in holding it together...We can use people like Ben...to help our cause and we do and he is a very effective operator in that position.

(Senior Associate, Sports Architecture firm, Sports Facility megaproject).

These quotes reveal that within the megaprojects, certain individuals played a key role in influencing the decision-making process. They were instrumental in effectively communicating key messages, supporting specific causes, understanding team dynamics, and initiating necessary changes. These individuals were noted for their exceptional ability at "getting things done". Project team members leveraged this knowledge to identify and engage with these effective operators, using their skills and influence, to steer situations in a direction beneficial to the project's success.

Destructive deviance

Contrasting with constructive deviance, destructive deviance encompasses behaviours that not only deviate from the established standards of megaprojects but also result in negative outcomes. In the case studies, tactics such as foot dragging and disrupting hierarchy were identified as forms of destructive deviance. These tactics had detrimental effects on the projects.

Box 4.11 Case study – exploiting ambiguity in rules for self-interest.

In pursuit of personal gains, some team members from the case study megaprojects engaged in deceptive and opportunistic behaviours. This practice of destructive deviance starkly differs from the goal-oriented, positive approach of constructive deviance. Instead of aiming to benefit the project, those employing destructive deviance sought to manipulate situations for their own benefit, often driven by hidden agendas and self-interest.

Project team members acknowledged that successful solutions in megaprojects often extend beyond the application of standard formulas. They perceived the process as a contest where each player had an equitable opportunity to succeed. This perspective was particularly evident in the stories from the airport case study megaproject, where the "games" played by the contractor were a recurring theme. The stories frequently referred to the diverse tactics, sometimes described as "gimmicks".

This approach was grounded in the understanding that any gap or ambiguity in the rules of the project presented an opportunity for exploitation.

Contractors, in particular, were seen as adept at identifying and utilising these openings in the rule system to their advantage.

Throughout the stories shared, project team members displayed a keen awareness of the social structure within their megaprojects, particularly in terms of their own positions, roles, and ability to influence outcomes. Those who found themselves in positions of limited authority often sought creative ways to exercise power, even in circumstances where their ability to exert control seemed minimal.

In the Airport megaproject, "foot dragging" was used by the contractor to assert some degree of control in situations where they otherwise felt powerless. This strategy involved intentionally slowing down earlier phases of the project. By doing so, the contractor aimed to narrow the range of decision-making options available to the client as the project advanced, effectively using delay as a mechanism to influence the project's trajectory in a negative manner.

Outline of the governmentality framework for megaproject decision-making

The governmentality framework for megaproject decision-making is developed based upon a balanced perspective that the two key factors of governance and governing work in concert with each other. This framework is depicted in Figure 4.4. Here, the interplay between the formal and informal elements of governance and governing is explored, leading to the identification of four distinct behavioural contexts.

The vertical axis in the framework, representing governing, categorises deviant practices within megaproject decision-making as either constructive or destructive. When governing is at a high level, it indicates constructive deviance, suggesting positive deviations from the norm that contribute to project success. Conversely, low levels of governing are associated with destructive deviance, implying harmful or counterproductive deviations.

Governing primarily focuses on the social dynamics within the decision-making environment of megaprojects. This includes the informal relationships, negotiations, and tactics that are a part of everyday deviant practices. It examines how members of the project team interpret and implement the set governance frameworks and mechanisms at a practical or instrumental level.

This concept is crucial in understanding the inherent complexities and deviations that are a natural part of managing megaprojects. It acknowledges the autonomy granted to individuals and teams, and how this autonomy influences their actions and decisions, whether for personal gain or the collective interest of the project.

The horizontal axis in the framework, labelled as governance, represents the degree of legitimate authority exerted through various formal structures such as rules, regulations, and systems. These structures are primarily designed to manage

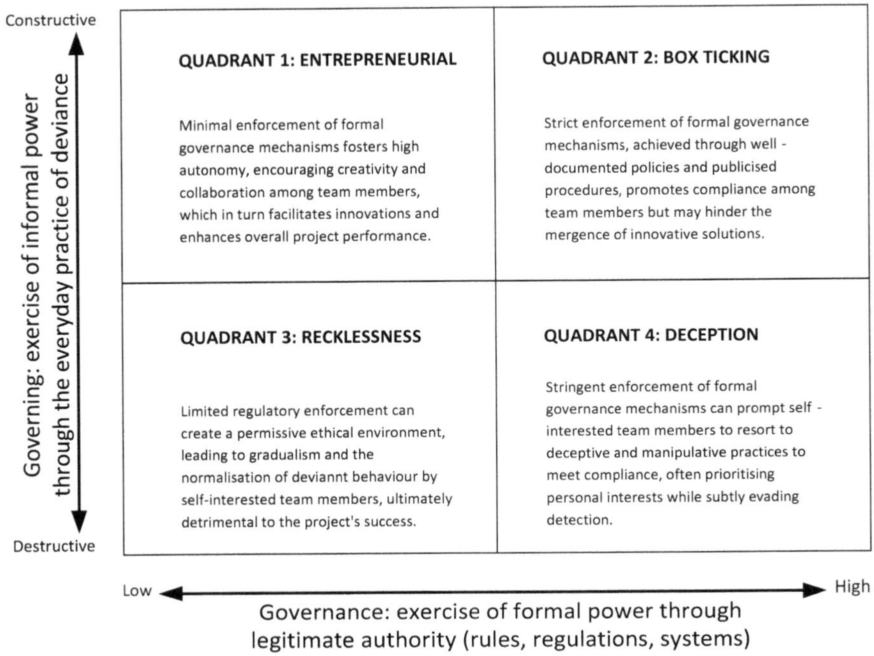

Figure 4.4 The governmentality framework for megaproject decision-making.

and control decision-making accountability. They also serve to define and maintain the roles and boundaries of all stakeholders involved in a project.

The aspect of governance is centred on regulation, the establishment of rules, and the creation of control arenas. This is done through adherence to established, formalised systems. Such systems include governmental regulations and explicitly prescribed frameworks and mechanisms for project governance. In essence, governance on this axis represents the structured, official side of managing a project, ensuring that decisions and actions align with predefined standards and legal requirements.

In the governmentality framework for megaproject decision-making, the four behavioural contexts are represented as quadrants. Each quadrant is defined by a unique combination of the governance and governing factors. These quadrants not only outline theoretical permutations of these factors but are also enriched with illustrative stories to provide real-world context and clarity.

Quadrant 1: Entrepreneurial

Quadrant 1 in the governmentality framework, referred to as the "entrepreneurial" quadrant, creates an optimal environment for decision-making in megaprojects. This quadrant is characterised by a unique balance where constructive deviance is not just allowed but actively encouraged. This encouragement of constructive deviance promotes innovation and adaptability within the project team.

Simultaneously, there is a more flexible approach to the enforcement of rules, regulations, and systems. Instead of rigidly adhering to these structures, there is room for interpretation and adaptation, allowing for dynamic and responsive decision-making. This flexibility does not mean the abandonment of governance but rather its strategic application to foster a more creative and effective project environment.

In essence, Quadrant 1 combines the best of both worlds: it maintains a certain level of formal governance to ensure accountability and order, while also promoting a governing style that values creativity, adaptability, and constructive deviation from the norm. This combination is seen as ideal for fostering a thriving and successful environment for megaproject decision-making.

Box 4.12 Entrepreneurial quadrant in action.

Constructive deviance, as illustrated in the entrepreneurial quadrant of the governmentality framework, is recognised as a vital source of creativity that can significantly enhance problem-solving and overall performance in megaprojects. This type of deviance involves employing various tactics such as gatekeeping and disrupting hierarchy. Such actions often represent a departure from standard rules and protocols. However, the strategic use of these tactics is crucial for the smooth functioning of projects, contributing to timely completion and fostering innovative or creative solutions.

These tactics are not employed with the intent of undermining rules or being deceitful. Rather, they are acts of initiative taken by project team members to advance and benefit the project. Project team members often find themselves in situations where adhering strictly to formalised processes and procedures may not yield the desired results. In such cases, they opt to reshape situations and outcomes using alternative means, which may sometimes involve bending or even breaking the rules.

The stories from project team members highlight that everyday acts of deviance are not exercises in rule-breaking for their own sake. Instead, they are pragmatic decisions made in the interest of project progress and success. Despite the apparent violation of formally prescribed protocols, this behaviour can be viewed as functional, contributing towards achieving project objectives.

The successful implementation of constructive deviance in megaprojects, as exemplified in Quadrant 1, hinges on granting a high level of autonomy to project team members. This autonomy is crucial for fostering an entrepreneurial spirit necessary for resolving project issues. Project team members in such environments are often strategic in their use of the system, discerning when to exercise their autonomy to deviate from standard protocols. Their decision-making about engaging in informal discussions (backdoor conversations) or opting for more formal escalation of project matters is nuanced, deliberate, and context-dependent. This strategic and selective application of constructive deviance tactics is not merely manipulative, it's a calibrated approach to maximise project benefits.

Project team members utilise a range of roles and resources available to them, selectively drawing from these to influence the direction of power in social interactions. While individuals may not have complete freedom in choosing their roles in every situation, they do possess a repertoire of roles and resources from which they can strategically draw.

In this quadrant, the role of the client is particularly significant. As change agents, clients are instrumental in creating and sustaining an environment that supports and encourages innovative behaviour. By setting up a context conducive to creativity and flexibility, clients play a pivotal role in enabling the kind of autonomy that is essential for the effective practice of constructive deviance in megaprojects.

Box 4.13 Case study – entrepreneurial negotiation for innovation.

The practice of constructive deviance in the context of "entrepreneurial" is exemplified by the example involving the use of a novel cooling system for the National Stadium in the Sports Facility megaproject. This case illustrates how project team members employed the tactic of "swapping deals" to gain regulatory approval for an innovative and sustainable bowl cooling system.

Initially, the regulatory body was discouraging the spillage of air conditioning from indoor spaces to outdoors, a practice viewed as unsustainable. The introduction of the bowl cooling system posed a challenge, as it could be perceived as setting a precedent for outdoor air conditioning, seemingly contradicting the regulatory body's stance. However, the local government agencies carry social expectations to be seen as "doing the right thing" in the eyes of the public.

In this situation, the design team, understanding the complex dynamics at play, was willing to engage in a "give-and-take" approach. They worked collaboratively with the consortium to devise a balanced solution, one that would compensate or offset the government authority appropriately. This process required bypassing standard hierarchical channels and directly engaging with agency directors who had the strategic insight to address the issue.

The critical role of the client in this scenario cannot be overstated. Given their position of power and influence, the client was instrumental in facilitating the approval process. This example emphasises the necessity of deviating from standard procedures in certain instances. By engaging in swapping deals, the project team members were able to navigate the rigid regulatory environment, securing approval for beneficial and innovative project components.

This case study highlights the nuanced application of constructive deviance within a highly regulated environment. It demonstrates how strategic flexibility and negotiation can lead to optimal outcomes, balancing compliance with innovative solutions in megaproject decision-making.

Quadrant 2: Box ticking

Quadrant 2, labelled as the "box ticking" quadrant, represents a scenario where there is a strong enforcement of rules, regulations, or systems, coexisting with the practice of constructive deviance. In this quadrant, the role of the client and regulatory authorities is crucial in shaping the ethical climate of megaprojects. They enforce governance through tightly controlled mechanisms or systems, such as thorough documentation and well-publicised channels for compliance. This leads to the creation of an anti-permissive ethical climate, which is essential for ensuring compliant behaviour.

However, this anti-permissive environment, while pivotal for compliance, may not always be conducive to fostering creativity and innovation. Strict adherence to rules and regulations, or simply "ticking the box" can sometimes be counterproductive. In such contexts, deviant behaviour, if practised constructively can be more beneficial.

Box 4.14 Case study – balancing innovation and compliance.

The introduction of the innovative bowl cooling system in the Sports Facility megaproject faced significant challenges due to conflicting priority agendas among key stakeholders. This situation highlights a common dilemma in megaprojects: the tension between innovation and regulatory compliance. Innovations, by their very nature, represent a departure from the norm and often do not align neatly with existing codes or regulations.

Megaprojects are characterised by their incorporation of novel solutions, making adhering to established codes and achieving regulatory approval a significant hurdle. Despite the system's efficiency significantly surpassing the basic requirements of building regulations and codes, its novel nature placed it at odds with existing regulatory frameworks.

This scenario highlights the intricate balancing act required in megaprojects between pushing the boundaries of innovation and navigating the complex web of regulatory requirements. The design team's challenge lay not only in proving the technical superiority of the bowl cooling system but also in convincing the authorities of its viability within the context of existing regulations.

Megaprojects, with their unique characteristics and high-profile nature, inherently provide a rich environment for implementing innovations. Strategic decisions in megaprojects are driven by interests and motivations that are typically absent in more conventional projects. These decisions often require thinking beyond standard approaches, leading to innovative behaviour that may not align with established protocols. The case of the bowl cooling system illustrates how innovation can clash with the rigidity of rules and regulations, such as those enforced by the regulatory body. In this instance, the stringent regulations posed a significant challenge, not only prolonging the compliance process but also stifling creativity and restricting innovation.

Despite these challenges, there was a marked determination within the client network. This resolve emphasises the unique commitment to innovation often found in megaproject environments. However, it is important to note that this level of dedication to innovation may vary from project to project. In some cases, the challenges of navigating strict regulatory landscapes and the associated risks might deter stakeholders from pursuing innovative solutions.

A decision-making environment characterised by a high level of enforcement of rules and regulations may not be the most conducive for fostering innovations. Such an environment could potentially discourage project teams from engaging in groundbreaking behaviour. This points towards the importance of acknowledging and even embracing a certain degree of deviant behaviour, provided it is constructive in nature, to support creativity and innovation in megaproject environments.

Quadrant 3: Recklessness

Quadrant 3 in the governmentality framework, termed as recklessness, is characterised by a combination of low governance and the practice of destructive deviance. This quadrant illustrates a scenario where the decision-making environment is marked by lax enforcement of rules and regulations, and where self-interested project team members engage in destructive deviant behaviours.

While high autonomy, as discussed in Quadrant 1, can be instrumental in fostering creativity and innovation, in the context of Quadrant 3, this autonomy, when paired with destructive deviance, can have detrimental effects. In an environment where rules and regulations are loosely enforced, and questionable practices are tolerated or even overlooked by those in power, there is a risk of these practices becoming increasingly normalised. When destructive behaviours are not adequately checked or discouraged, they can gradually be accepted as standard operating procedures. Quadrant 3 represents a cautionary scenario where the absence of strong governance and the presence of destructive deviance can lead to a toxic and potentially damaging project environment.

Box 4.15 Case study – destructive deviance and its consequences.

The scenario described in the Airport megaproject highlights the detrimental effects of reckless behaviour in the context of megaprojects. In this case study, the contractor exhibited what can be termed as "gamesmanship", deliberately ignoring established lines of authority and procrastinating on tasks when it was advantageous for their own goals. This kind of behaviour is a classic example of destructive deviance where actions are intentionally and purposefully designed to influence the decision-making process in favour of personal objectives, with little or no consideration for the overall performance of the project.

These actions were not accidental or incidental; they were strategic and calculated, aimed at manipulating outcomes to benefit specific parties. A key issue in this scenario is the lack of negative consequences for the contractor's actions. Without any significant repercussions, the contractor was incentivised to continue practising destructive deviance, eventually establishing it as the new standard in the project environment.

Despite the destructive deviance demonstrated by the contractor, there was a belief among the project team members in the contractor's capacity and competency. They perceived the contractor as not only adept at navigating the complex web of rules and regulations but also skilled at transforming a highly constrained environment into one ripe with opportunities for achieving their own objectives.

Quadrant 3 represents a highly risky environment in the context of megaproject decision-making. The potential consequences of destructive behaviour necessitate vigilant oversight and intervention by those in positions of authority. One of the essential aspects to guard against this quadrant is the process of gradualism – the slow and often unnoticed acceptance and normalisation of destructive behaviours. Without careful oversight, there is a risk that such behaviours, initially viewed as aberrations, can gradually become the norm. This normalisation can have far-reaching and damaging consequences for the project, including compromised integrity, reduced efficiency, and potential legal and ethical violations. Therefore, it is imperative for those in leadership and governance roles to take proactive steps to prevent this gradual acceptance of destructive deviance.

Quadrant 4: Deception

Quadrant 4, identified as deception, presents a unique intersection of high governance with the practice of destructive deviance. In this environment, a strict, rules-based, and tightly controlled project setting exists, ostensibly limiting actions and behaviours. However, paradoxically, this restrictive environment can also create opportunities for individuals to exercise control, even in situations where they otherwise might feel powerless.

The essence of this quadrant lies in the way rules and regulations can inadvertently create both opportunities for resistance and the means to resist. While rules are designed to govern and control, they can also, in their specificity and complexity, give rise to areas that are less clearly defined or regulated. Individuals who are adept at navigating these environments can exploit these gaps for their own ends, often engaging in practices that are at odds with the intended purpose of the rules and the overarching goals of the project.

Box 4.16 Case study – navigating compliance through covert deception.

The example from the Airport megaproject (detailed in Chapter 7) illustrates a striking instance of destructive deviance in the context of Quadrant 4. The contractor's actions, characterised by foot dragging, exemplifies a disregard for the possible disastrous consequences that such actions could have on the project. Instead of outright refusal to undertake work, this foot dragging tactic was employed as a subtle assertion of autonomy within the framework of compliance. This behaviour allowed the contractor to "slip under the radar", as it did not involve openly breaking any rules. Yet it effectively manipulated the governance mechanisms, conforming to them superficially while essentially seeking to escape their intended control.

This approach to destructive deviance highlights how project team members can remain unnoticed while they navigate through short-term tactics that deceptively demonstrate compliance in megaprojects. The contractor's familiarity with the bureaucratic system, and their adeptness at using different deceptive tactics to suit various situational needs, was evident in the stories recounted. These narratives showcase the ease with which the contractor manoeuvred within the system, using a blend of compliance and manipulation to achieve desired outcomes.

The contrast between Quadrant 2 and the scenario depicted in Quadrant 4 provides a deeper understanding of the implications of high governance in megaproject environments. In Quadrant 2, high governance leads to what can be described as blind conformance to rules and regulations, often resulting in a stifling of creativity and innovation, as project team members focus more on adhering to rules rather than exploring new and potentially more effective solutions.

Conversely, Quadrant 4 presents a different challenge associated with high governance. Here, the strict rules and regulatory frameworks do not necessarily ensure genuine compliance. Instead, they may inadvertently encourage deceptive behaviour. In this quadrant, project team members may appear to comply with rules on the surface but engage in manipulative or deceptive practices behind the scenes. The key takeaway from this is that high governance, while essential for order and accountability in megaprojects, is not a panacea.

Box 4.17 Case study – subverting intent for deceptive compliance.

The example from the airport case study (detailed in Chapter 7) vividly illustrates the use of "rule literalness" tactics as a form of deceptive compliance. In this case, the contractor employed a strategy where the purpose of a rule was subverted through its rigid observance, highlighting a sophisticated understanding of the power dynamics at play within the framework of rules and regulations.

The contractor, understanding the contingent and open-ended nature of the design process, recognised an opportunity to manipulate the situation to their advantage. By repeatedly emphasising that all their actions were based on formal approvals provided by the client on the structural drawings, the contractor used the "rule literalness" tactic to deflect blame. The assertion was that since everything was formally approved, their firm could not be at fault.

This tactic leverages the authority and stability that written approvals and documentation provide. Textual authority, while crucial for clarity and accountability, can also become inflexible, making it challenging to revise or contest decisions once they are recorded. The contractor capitalised on this aspect, using the durability and unchangeability of the client's recorded approval at one point in time to justify their actions at a later stage. By rigidly adhering to the approved documentation, the contractor claimed to be merely following the rules, thus absolving themselves of any wrongdoing.

However, the situation called for a more adaptive approach. To achieve the desired and consistent ceiling height, it was necessary to revise or adapt the approved documentation to meet the evolving needs of the project. In this context, the contractor used rule literalness not as a means to uphold the integrity of the project, but as a tool to manipulate the situation to their advantage, revealing a complex interplay between power, rules, and compliance in megaproject decision-making.

Chapter summary

This chapter introduced a governmentality framework for megaproject decision-making, emphasising the need for a blended approach that incorporates both formal governance mechanisms and informal governing practices. Through two megaprojects, the chapter explored the tension between formal project frameworks and the practical realities of decision-making in megaprojects. The concept of deviance was dissected, illustrating how project team members employ a range of tactics for various objectives, which can lead to both positive and negative outcomes. The chapter also described four quadrants of the governmentality framework – entrepreneurial, box ticking, recklessness, and deception – each representing a unique interplay of governance and governing. The chapter has highlighted the complexities of megaproject decision-making and the need for a governmentality framework that accommodates this complexity.

References

Beckert, J. (1999). Agency, entrepreneurs and institutional change: The role of strategic choice and institutionalised practices in organizations. *Organisation Studies*, 20, 777–799.

Biesenthal, C., Clegg, S.D., Mahalingam, A., & Sankaran, S. (2018). Applying institutional theories to managing megaprojects. *International Journal of Project Management*, 36(1), 43–54.

Brislin, R. (1991). The art of getting things done: a practical guide to the use of power. New York: Praeger Publishers.

Ewick, P., & Silbey, S. (2003). Narrating social structure: Stories of resistance to legal authority. *American Journal of Sociology*, 108(6), 1328–1372.

Marshall, N. (2006). Understanding power in project settings. In D. Hodgson, & S. Cicmil (Eds.), *Making projects critical* (pp. 207–231). London: Red Globe Press.

Ninan, J., Mahalingam, A., & Clegg, S. (2019). External stakeholder management strategies and resources in megaprojects: An organizational power perspective. *Project Management Journal*, 50(6), 625–640.

Pinto, J.K. (1996). *Power and politics in project management.* Newtown Square, PA: Project Management Institute.

Siva J., & London, K. (2016). Disrupting social structure to achieve innovation on Public-Private Partnership megaprojects: a narrative analysis of instruments of power in the Singapore Sports Hub case study, In *New Forms of Procurement PPP and Relational Contracting in the 21st Century*, Routledge, Abingdon, 193–215.

5 Box ticking

Blind conformance to rules could be counterproductive

Box ticking and compliance: is this really good behaviour?

In megaproject decision-making, the seemingly straightforward path of box ticking and achieving compliance raises an important question: Is this path synonymous with "good" behaviour, or does it merely scratch the surface of ethical engagement?

Quadrant 2 of the governmentality framework, known as "box ticking", combines a strict adherence to rules, regulations, or systems with the practice of constructive deviance. In megaprojects, both the client and regulatory authorities play pivotal roles in shaping the ethical climate. When governance is enforced rigorously through tightly controlled mechanisms or systems (such as detailed documentation and widely publicised reporting channels), it can create an environment where ethical compliance is strictly upheld. However, this kind of environment, while crucial for ensuring compliance, may not always be the most conducive to fostering creativity and innovation.

Sometimes, blindly following rules and regulations or simply going through the motions of "ticking the box" can hinder progress. In certain situations, constructive deviant behaviour, when channelled positively, may prove to be more beneficial. It is about recognising that there are instances where deviating from the norm can lead to innovative solutions and improved outcomes.

This chapter tells the story of the bowl cooling system innovation, which serves as a striking illustration of the pitfalls of the "box ticking" mentality. It outlines the critical events in the journey of the bowl cooling system innovation, highlighting a significant challenge that emerged, stemming from the regulatory issues involving the governing body for construction regulations in the country.

The regulatory body initially found that the proposed system violated their regulations. In response, the project team embarked on an in-depth process of computer modelling and working out calculations to determine the most efficient way to ensure spectator comfort. Despite their efforts, there was a dilemma: the team believed that their innovative approach was the best solution, but it faced resistance from the authorities.

Interestingly, there was an alternative approach suggested, one that would consume "four times the energy" but might be more palatable to the authorities due to its avoidance of potential public relations challenges. Eventually, after

DOI: 10.1201/9781003377481-5

deliberations, the regulatory body granted regulatory approval for the innovative cooling system, with the condition that the system's energy usage must be fully offset by renewable energy sources. This story highlights the complexities of balancing innovation, compliance, and power dynamics in megaproject decision-making.

The upcoming sections delve into the development and implementation of the bowl cooling system, shedding light on the pivotal moment when the innovation was initially perceived as a deviation from the norm. It also provides insight into how the project team responded to this deviation. The story of the bowl cooling systems serves as a potent reminder that simply adhering to established rules and protocols may not be sufficient to overcome challenges, emphasising the need for alternative approaches.

This narrative also highlights a compelling aspect of megaproject dynamics: the exercise of political power can sometimes supersede rational decision-making. Despite the design team's expertise and their presentation of an innovative and efficient system, they found their professional influence to be less effective than the client's legitimate power derived from their position within the network of government agencies. Ultimately, it was through the use of deviant tactics, such as disrupting hierarchy and swapping deals, that the system achieved regulatory approval.

Proposal of the bowl cooling system innovation

The bowl cooling system was proposed by the consortium during the preferred bidder stage in response to the requirements outlined by the government agency client. The primary focus was on enhancing spectator comfort, as reflected in the Invitation to Tender, which outlined that the project should fulfil the requirements of event promoters comprehensively while offering spectators and participants an experience that is affordable, comfortable, and enjoyable. It also specified that the PPP contractor is tasked with ensuring comfortable operational conditions for all users on event days and that environmental systems must be integrated as necessary to maintain conditions that are conducive and comfortable for all users.

In response to this, the bowl cooling system was incorporated into the client's final design brief, aligning with the consortium's contractual commitment to the client. Their goal was to achieve a GreenMark GoldPlus award, which, at that time, represented the second highest level within the country's system for evaluating environmental performance of buildings.

Design strategy: sustainable semi-outdoor stadium design

The design strategy of the cooling system was driven by two key factors: the architectural vision of creating a semi-outdoor stadium and a strong commitment to sustainable stadium design. Consequently, the conventional approach of installing air conditioning in a fully enclosed stadium was not considered. Instead, preference was given to passive measures to meet the required criteria. As a result, the concept of the bowl cooling system was proposed, relying significantly on enhancing air movement. This innovative system was also designed to minimise energy use by focusing on treating the microclimate immediately surrounding the spectators rather than the entire volume of the stadium.

A combination of fixed and retractable roofs was used in the stadium's design. The fixed roof was strategically positioned to offer shade to the spectators while allowing sunlight to reach the playing field through a sizable opening. On the evening prior to an event, the retractable roof would be closed not only to provide shade during the event but also to prevent the stadium's reinforced concrete structure from absorbing excess solar energy.

Despite the stadium being designed as a non-enclosed structure, it became evident that relying solely on natural wind movement would not suffice to maintain the desired comfort level. As a solution, the "bowl cooling" scheme was introduced to meet the comfort criteria effectively. To conserve energy, the scheme featured relatively small air handling unit (AHU) zones, with only the occupied areas being activated.

The scheme underwent extensive refinement through collaboration between the design and construction teams, involving several revisions to address various requirements. In addition to computer modelling conducted by the design team, full-scale mock-ups of seating areas were constructed to closely assess and monitor the system's performance.

Is good practice not good enough?

After the finalisation of the project agreement, an issue emerged with the bowl cooling system. This concern was flagged by the regulatory body, as illustrated in Box 5.1.

Box 5.1 Lesson – the importance of considering diverse forms of power.

The cooling system was spoken about in terms of being something that was a "violation of regulation" and a "concern" which was "going off the rails". The path taken by the project team was fraught with challenges, demanding substantial effort and dedication from the team. The quotes below show that it wasn't just about who had the best idea; other kinds of influence also got in the way. These other forces basically put up roadblocks, making it hard for the project team to get their well-thought-out proposal accepted.

> So, in the last 2 years a lot of our energy actually has gone into constructive presentations with the regulatory body to show that our solution is good practice.
>
> (Senior Associate, Sports Architecture Firm, Sports facility megaproject)

> So, we developed that over several period of years with the client… and you know, as soon as you work with government! So, we had instructions from one part of government – and that necessarily creating a PR problem for another part of government.
>
> (Senior Design Manager, Design and Build Contractor, Sports facility megaproject)

In many Asian countries, it is not unusual for establishments like restaurants to extend their air conditioning from indoor spaces to nearby outdoor areas. This practice offers patrons the unique experience of being outside while still enjoying climate-controlled comfort. However, the regulatory body has been working to put an end to this energy-inefficient method for years, using stricter regulations and enforcement measures. They had also signalled that they might bring in additional rules to further limit this practice. Given this context, there was a period of uncertainty regarding whether the national stadium could use its "bowl cooling system", especially since the regulatory body viewed the system as unacceptable. As the design manager put it, the situation was "up in the air".

The bowl cooling system is not exactly air conditioning, but the regulatory body expressed concern that the public might see it that way. The key difference is the bowl cooling system does not control humidity; it mainly relies on air movement for cooling. Despite this, the regulatory body feared that giving the green light to such a system could set a precedent. Specifically, it might encourage the use of actual air conditioning in outdoor areas, which would work against the regulatory body's ongoing efforts to curb such environmentally unfriendly practices.

When logic and technical merits fall short of broader power dynamics

For an extended period, the design team was in a position of having to justify the bowl cooling system to the regulatory body. Their defence mainly focused on the technical efficiency of the system. The level of frustration experienced by the design team during this period is evident in the quotes that follow.

Box 5.2 Lesson – a singular emphasis on logical and technical merits, while adhering to compliance measures, overlooks the broader power dynamics that can influence the outcome.

The design team faced a roadblock, not due to a lack of technical rigour but because of broader considerations. Despite utilising extensive computer modelling to show that their bowl cooling system was the most energy-efficient solution for spectator comfort, they felt stymied by regulatory constraints. The team believed that the regulatory body might have preferred a far less innovative approach – one that could consume up to four times as much energy – simply because it would not present the same public relations challenges.

> One of our arguments…we could've arrived at a solution which would've used…2 or 3 times as much energy…but that generally wasn't accepted because to the authority they're still looking at us

documenting the project where we're bringing cool air into an outside space

(Senior Associate, Sports Architecture Design Firm, Sports facility megaproject)

What we're doing is the most efficient thing but from their point of view, not to hinder this other law, it'll be better if we built an enclosed dome and used four times the energy…we didn't anticipate this kind of problem or this political wrinkle that you know, kind of cast doubt on what we have done.

(Senior Design Manager, Design and Build Contractor, Sports facility megaproject)

they also don't want to get into a situation that there is criticism… There are comparisons to other projects where semi outdoor spaces have been cooled and the regulatory body hasn't allowed that and that is part of the reason why this has become a sensitive issue

(Senior Associate, Sports Architecture Design Firm, Sports facility megaproject)

It flies in the face that something that the regulatory body is trying to restrict

(Senior Design Manager, Design and Build Contractor, Sports facility megaproject).

The regulatory body had been working for years to discourage the cooling of outdoor spaces, so the bowl cooling system was viewed as directly contradicting their efforts. Although the regulatory body recognised that the system was not traditional air conditioning, they were concerned about public perception. They feared that if people mistakenly thought it was air conditioning, it could undermine their authority and the rules they enforce for construction in the country. In this situation, strictly following guidelines and prioritising energy efficiency did not help the project, despite the design team's good intentions. This led to a dilemma where deviant tactics had to be employed to secure regulatory approval for the cooling system.

Chapter summary

Megaprojects are not just about scale; they are about pushing boundaries. Megaprojects often call for innovative ideas and solutions not commonly seen in smaller projects. However, innovation often clashes with the rigid rules and protocols that govern such projects. The bowl cooling system innovation serves as an example:

the regulatory body's strict regulations made compliance a long and stifling process, limiting the creative solutions the design team wanted to implement.

Not all projects are as committed to innovation as this one was. Teams might shy away from groundbreaking ideas if they anticipate a maze of red tape. This suggests that a more flexible approach to regulations could promote creativity. Allowing for a constructive device, that is, rule breaking when it serves a greater good, could foster a more innovative project environment.

6 Deception

Opportunistic use of power can result in beneficial or detrimental outcomes

Deception: beneficial or detrimental?

Deception is a multi-faceted concept in megaproject decision-making. It can be beneficial when it involves creatively navigating red tape that stifles progress. On the other hand, it can have detrimental effects if it means compromising ethical standards or manipulating stakeholders for personal interests.

Quadrant 4, also known as deception, illustrates a unique intersection where stringent governance intersects with acts of destructive deviance. Here, the existence of comprehensive rules paradoxically opens up avenues for individuals to exert a degree of control, even in highly regulated environments. It acknowledges that within highly regulated environments, there exist gaps – lacunae – that savvy individuals or groups can exploit. These gaps may arise due to ambiguities, or contradictions within rule systems.

While such exploitation can lead to detrimental outcomes, the ability to navigate these spaces is not necessarily malicious. Indeed, this capability can be crucial for innovation and progress, as strict rule adherence can sometimes stifle growth and adaptation, as illustrated in Chapter 5. Therefore, understanding this quadrant involves recognising the dual nature of rules in both enabling and disabling deceptive behaviour. This deceptive behaviour can manifest in various forms, ranging from benign rule-bending for the sake of efficiency or innovation to more malignant forms of deceit and manipulation for personal gain at the expense of the collective good.

This chapter presents two stories which exemplify how deception in megaproject decision-making can result in varied outcomes.

1 *Beneficial outcomes of deception*: The first story detailing the Plaza canopy conundrum illustrates the artful navigation of conflicting priorities, showing how divergent agendas can be aligned through adept project management and deceptive behaviour.
2. *Detrimental outcomes of deception*: The second story is a more cautionary account: the saga of an airport terminal. Here, the harmonious resolution found in the Plaza was elusive, as attempts to reconcile differing objectives fell short. The central issue lay in the terminal's ceiling height – a feature critical to both

DOI: 10.1201/9781003377481-6

the building's functionality and aesthetic appeal. Not only was the ceiling height of the terminal building compromised, but the debacle also resulted in increased time and costs to the project schedule and budget. This appeared to be doubly a problem because the usual benefits associated with reduced building ceiling height, cutting spending, and minimising time were not achieved.

These two stories highlight critical lessons for understanding the dichotomy of deception as both a facilitator of progress and a potential catalyst for ethical dilemmas and project failures.

The Plaza canopy conundrum

The Mass Rapid Transit (MRT) is one of the main modes of transportation in the country and the Stadium MRT station emerges as a vital hub, particularly for those heading to the Sports Facility megaproject. Named in homage to the National Stadium, the station not only serves the sporting complex but also holds a historical precedence over it, having been part of the urban landscape well before the Sports Facility's inception.

The Stadium MRT station is more than a mere waypoint for commuters – it's a landmark of design excellence, celebrated with the prestigious Design of the Year accolade at the President's Design Award in 2010. Their intent was to recreate that sense of awe and grandiosity, transforming a daily commute into an encounter with architectural splendour.

Since its opening, the Stadium MRT station has commanded recognition as a significant architectural entity, variously referred to by some as a "strong architectural space", "a very powerful kind of box", "a very sculptural object", and "a well-executed piece of work".

Box 6.1 Case study – the Plaza canopy conundrum

The Plaza canopy conundrum stands out as a classic case of project deviation. The task was seemingly straightforward: create a seamless connection between an existing MRT station and the new Plaza. Yet, the solution proved elusive. The journey towards the eventual canopy configuration unfolded with a series of challenges that some insiders likened to playful children's games, while others drew parallels to a persistent back-and-forth akin to table tennis matches.

At the heart of the problem lay a clash of perspectives. Each stakeholder brought to the table a distinct worldview, with the divergence in perspectives sparking tensions over what should take precedence. From the client's viewpoint, practicality took centre stage – the canopy was envisioned as a shield for the public against the inclement weather. They advocated for a substantial structure that would offer uninterrupted shelter for commuters as they transitioned from the MRT station to the Plaza.

Meanwhile, the design team's vision was driven by an artistic aspiration, where aesthetic purity held the reins. They championed the architectural sanctity of the existing station; its historical and aesthetic legacy was something the designers were committed to preserving.

The team faced a delicate balancing act: how to execute a design that embraced the functionality demanded by the client while honouring the architectural narrative they deemed integral.

> So, if you look at that station, it's a very clean form in itself, so it's a bit of architecture. In fact, you don't want to touch it anymore. It's a very sculptural object and it's a clean object. And it's a well-executed piece of work and if we go and stick a big canopy up against it... it will take away from that...but from a client point of view...they have other things they need to weigh up, ease of access and the functionality issues.
>
> (Senior Associate, Sports Architecture firm, Sports Facility megaproject)

> People in general would want to be sheltered...so this was something we thought...There was some initial resistance where they [the design team] actually wanted to remove this [canopy] totally... They thought that it's not required...that it's not within the architectural language used.
>
> (Senior Manager and Head of Design and Technical, Sports Facilities Division, Sports Facility megaproject)

The project, by its nature, invited the scrutiny and input of multiple regulatory bodies – the Land and Transport Authority (LTA), Civil Defence Force and Urban Redevelopment Authority (URA) – each with its mandates and requirements. Additionally, the venue operator, a critical component of the public–private partnership (PPP) consortium, also wielded influence over the project's direction. As noted by the Senior Associate of the architectural firm involved, these circumstances rendered the process "a bit more tricky".

> I think in the beginning there were mixed reviews so some said OK, some said it may work. Others said, "why not you speak to LTA?" [laughs] So there was a lot of...to and fro because...the decision kind of overlaps between the agencies, that's where it's a bit more tricky.
>
> (Senior Associate, Architecture firm, Sports Facility megaproject)

Their testimony reflects the intricate balancing act required to synthesise these various demands into a coherent and feasible solution, one that would satisfy not just the aesthetic and functional aspirations of the project but also adhere to the stringent requirements laid out by the governing entities.

The Chessboard of megaproject decision-making

When examining the inner workings of managing megaprojects, a landscape emerges where power plays a crucial role. Here, the performance of a project is influenced not only by contractual documents but also by the strategic exercise of power.

Box 6.2 Case study – autonomy to play the game of megaproject decision-making.

Decision-making in megaprojects often resembles a strategic game governed by rules that are frequently adapted or manipulated to achieve specific objectives. The continuous evolution of this game-like scenario is driven by the unique complexity of megaprojects, which demands that rules be flexibly applied and sometimes even set aside. Unlike conventional building projects, megaprojects require a level of adaptability in their processes. This includes provisions for proposing alternative solutions that meet performance-based requirements, rather than strictly adhering to prescriptive standards.

Insights from the Airport megaproject highlight the significance of offering flexible options to project teams, particularly in gaining acceptance for performance-based solutions from regulatory bodies. Such flexibility is deemed essential in the context of megaprojects. The Chief Operations Officer from the megaproject emphasised this point, stating, "For us, we are a bit flexible…you need things that are new; you cannot be too rigid". This reflects a client-side inclination towards solutions that may diverge from standardised practices, provided they demonstrate compliance with the necessary performance benchmarks. Validation of these alternative approaches can be achieved through diverse means, including performance tests, case studies, and evaluation reports.

It was further noted that prescriptive codes are frequently supplemented with performance-based standards. This dual approach accommodates the dynamic needs of megaprojects, allowing for compliance to be achieved through various methodologies.

> The good thing…is there's always a set of what you call prescriptive codes [but] there's also what you call performance based. So at least there's an avenue to allow you to look at it from a different point of view but still addressing some of these concerns that need addressing.
>
> (Senior Associate, Architecture firm, Sports Facility megaproject)

> What we did was we include in that the contractor can propose the alternative design if they think that their design is far better than the consultant design…it depends on the creativity of the project manager and the person.
>
> (Chief Operations Officer, airport operator, Airport megaproject)

> … some of their [client] statements of concept are very general, not specific. So when it comes to the general then we streamline what we have on site. Then from there we develop whatever we have on site with the CDC, then our contractor designs accordingly [and] then construct.
>
> (Senior Design Manager and Quality Assurance & Quality Control Manager, Airport Client Office, Airport operator, Airport megaproject)

In this environment, project team members are afforded considerable autonomy and flexibility. This freedom enables them to not just adhere to established rules but also to creatively interpret and manipulate them to their advantage.

Each megaproject can thus be likened to a sophisticated game of chess. The pieces on this board are clients, contractors, architects, engineers, government authorities and representatives, and a host of interested parties, each with their own agenda. Every move is deliberate, strategies are carefully planned, and the power to persuade becomes as vital as the blueprint for a skyscraper.

The view of megaproject decision-making practice as a game is based on one where tools, resources, and tactics are accessible and deployed by players to best their opponents. While project team members often perceive formally prescribed rules and regulations as inflexible and limiting, they simultaneously acknowledge that strategic actions and deviant tactics can significantly impact project outcomes. This recognition suggests that, despite the constraints of formal rules, the creative and strategic manoeuvring of team members plays a crucial role in shaping the direction and success of a project. Therefore, the expertise and accumulated experiences of individuals involved in a megaproject are pivotal in determining the project's outcomes across a range of scenarios.

Project team members working on the Plaza canopy development acknowledged the unpredictability of project results and their own role in influencing these results. When managed appropriately, outcomes can be aligned to suit personal agendas, like in the case of the West Plaza.

Spheres of authority and stakeholders of the shadows

The concepts of "spheres of authority" (Donzelot, 1979) and "stakeholders of the shadows" (Winch, 2017) are useful for understanding how power operates within the network of professional enclosures and agencies. The example quotes from megaproject team members in Box 6.3 shed light on how relational power is distributed across these spheres. Interestingly, these spheres of authority often emerge from dynamic relations among various informal authorities who are not formal clients but play a significant role in shaping the ethical climate and general direction of the megaproject decision-making process. This, in turn, influences the participants' propensity to engage in deviant behaviour.

Box 6.3 Spheres of authority and stakeholders of the shadow.

The Technical Director from the design and build contractor firm from the Sports Facility megaproject pointed out the challenges the design team faced in trying to meet the diverse needs of different clients, both official and unofficial. He gave an example of how these different clients could have opposing demands regarding the same issue:

> We, the design team, we have to satisfy many clients…and we need to think "OK should I go to the URA first or should I go to the SPV first" … You have the consortium who may say the access to the retail commercial building is not good, I'm not going to make good revenue. And you have URA who is going to say, "oh this is, I don't like this in terms of the integration of the Sports facility to the other roads". So that's a good example of clients having different opinions on the same subject

Another example is provided by the Senior Project Architect from the Client Design Consultant (CDC) team from the Airport megaproject to highlight the dynamic nature of the external environment within which team members operate:

> For example, this particular arrangement for the customs screening… how many times ding dong, they go to the customs and we've been called by the customs. Ok we want this from the previous officer…all the confirmation from all this people [and] departments…We want like this, like that and then ding dong ding dong and then the JV architect comes up with drawings and they go with that and endorsed that, now the previous officer left and comes another officer… and says, "oh the government says that we have this, this, this"…OK boss, we want you to endorse this and he asks, "have you done my things? Have you followed my requirements?" Yes sir, everything is OK then OK, [he] signs it…

These quotes show the diverse thought processes and actions of these "authorities" across various contexts, distinctly separate from the lofty considerations of high-level megaproject governance mechanisms.

These observations align with and support existing research on the normalisation of deviance, such as the work of Hajikazemi et al. (2020). This body of research suggests that the practice of deviance is shaped not only by internal organisational factors but is also significantly influenced by the external environment in which organisations operate. This reveals the complex interplay between internal and external factors in defining the ethical and operational aspects of megaproject decision-making.

Navigating project communication: the art of timely persuasion

The importance of sharing information at the right time and in the right way was emphasised by the Senior Design Manager from the Design and Build Contractor. He noted how quickly minor issues could turn into complex political situations if not handled appropriately. He shared an example where the design team's idea for visually dividing the Plaza from the Stadium MRT station was initially opposed by the client because it was presented unexpectedly, which led to confusion and the issue becoming tangled with other matters.

Box 6.4 Case study – managing expectations and controlling information flow.

The Technical Director of the Design and Build Contractor emphasised that a great deal of time and effort is spent on navigating the system to get approvals from various parties. Due to the wide range of expectations from these parties, the design team invests heavily in planning to meet everyone's needs. Addressing the example of conflicting priorities, the technical director attributed the resolution to the team's ability to successfully convince those involved.

> The approval mechanism of the documents is something quite complicated… and we spend quite a lot of time to make the system work, of who is approving what…Yes there was no battle, it's just that people were convinced, that's all. And it takes time to convince people.
>
> (Technical Director, Design and Build Contractor, Sports Facility megaproject)

> Certainly, some conversations are held quietly to try and prepare to get someone else ready to be bringing the information out in the form which is going to work for everybody… but the people who are doing the work…they're not necessarily attuned to precisely the message that we want to hand over to the authority. So, we do a little bit of work in making sure that they become attuned to that.
>
> (Senior Associate, Sports Architecture Firm, Sports Facility megaproject)

The Senior Manager pointed out that the design team's provision of high-quality information to regulatory bodies in support of their proposals seemed to help streamline the process. The Senior Associate explained that to get the necessary information from different members of the design team, they often had discreet conversations. This was done to subtly influence the team to present the desired information to the authorities.

Influencing outcomes through strategic representation

In the final design, the team decided to add canopies on the sides of the station, rather than just one big canopy at the main entrance. This change was agreed upon by the client and met the standards of various regulatory bodies. The design team argued that this approach was crucial for managing crowds effectively. However, their main goal was to preserve the station's striking visual appeal, which they referred to as the "fantastic gesture view".

Box 6.5 Case study – cleverly advancing architectural vision through functionality-focused arguments.

In recognising megaproject practice as an arena of contestation, project team members believed they had a range of tools and resources at their disposal, enabling them to strategically reach their goals as necessary. Stories emerged detailing the design team's tactical sharing of technical details with the client, aimed at subtly advancing their underlying agenda: to create a visual separation between the Plaza and the existing MRT station. The design team strategically presented their argument focusing on crowd control. This was a clever tactic to fulfil their architectural goal without drawing attention to their intention.

> Using crowd control as a diversion to an aesthetic...we couldn't bear the look of this thing. And the argument of crowd control, it just happened to align with what we wanted aesthetically... Architects, we have an aesthetic desire or like doctors have a credibility and in this case – [our argument was that] it won't burn down.
>
> (Senior Design Manager, Design and Construct Contractor, Sports Facility megaproject)

> In the end we have resolved it and now people are happy with it but now the people are focussed on, it's fundamentally a wrong thing to do from a crowd control point of view. It's to bring people to the mouth of the stadium and never to allow the opportunity for a crowd crash.
>
> (Senior Project and Design Manager, Architecture firm, Sports Facility megaproject)

The design team, with their experience in crowd control design, used their knowledge to support their design preferences. For example, the fire authority's rule to keep the station exit open for smoke clearance in case of a fire aligned with the team's plans. Although their strategic deviation did not lead to any negative outcomes, the team did distort the project's requirements to suit their own interests.

The ceiling height debacle

While the story of the Plaza canopy shows how conflicting priorities were managed successfully, another story reveals a less effective compromise concerning the inconsistent ceiling height in a terminal building at an airport megaproject in another Asian country. In different parts of the terminal building, the ceiling height varied from 2.1 to 2.8 m, averaging 2.6 m. However, the initial project requirements given to the contractor stated that the minimum ceiling height should be 2.8 m.

Box 6.6 Case study – the ceiling height debacle.

The ceiling height issue occurred due to a lack of coordination among the structural, mechanical, and electrical engineers on the project. In complex megaprojects like this, people from various disciplines work together, often leading to different interests and conflicting priorities. Each team member, coming from different organisations, had their own goals and challenges. This sometimes led to them prioritising their own organisation's objectives over the project's common goals. The Senior Project Architect from the CDC team shed light on how this self-interest caused the varied ceiling heights in the terminal building:

> ...when the slab is thicker the cost is higher and they [contractors] don't like that. They would prefer the deeper beam with thinner slab... they said, "oh we [will] tackle other issues later on. Why must we think about that now? We have to settle the structural faster". OK so settled, then comes the M&E [mechanical and electrical services], "I couldn't be bothered about other things, I'll settle my things ... So I settle my part" ... then these guys also don't care...that's the problem with this and now we only have 2.4m +-...[which is] very low.
>
> (Senior Project Architect, Client Design Consultant – Architecture firm, Airport megaproject)

The inward-focused mindset of the contractor and the structural, mechanical, and electrical engineers was marked by short-term thinking and some deceit. They were more concerned with finishing their parts, even though they knew it could harm the overall project. Coordinating such a diverse group with different goals and specialities is challenging but crucial for success. In complex megaprojects like an airport, disagreements and varying approaches can negatively affect both the project and team dynamics. Team members often focus on their specific tasks rather than the overall project objectives. To protect their interests, they use various forms of power and strategies to achieve better outcomes for themselves.

Shifting blame: how rule literalness became a tactic in the ceiling height debacle

There was a clear need for the contractor and various specialists to collaborate on a solution that met everyone's needs without compromising the ceiling height. However, the contractor and engineering consultants pretended not to know about the activities of other teams involved in building the structure and installing mechanical and electrical services. By doing so, they saved time and effort by just focusing on their own areas, neglecting the need to work together with other trades to achieve the required ceiling height.

Box 6.7 Lesson – playing dumb can help to shift blame elsewhere.

The contractor acted as if they were just following the client's approval of the initial structural plans. This decision affected other parts of the project and resulted in inconsistent ceiling heights. The CDC architect blamed the contractor for poor coordination, causing the ceiling height problem, calling it "their fault". However, the contractor defended their actions, arguing it was a matter of debate. They pointed out that the client had approved the structural drawings, saying, "They approved our structural drawings":

> They try their best to get us to do [the changes] but there's a lot of things…They approved our structural drawings. The structure defines the ducting. …My slab to slab is 4.2, my beam is 750. So, 750 minus 4250 is 3334 so below the beam I need to have my duct and my duct is 450. Then we have firefighting and…we have no space…Everything was approved by them [the client]! All these issues, some part of it they have to swallow and some part of it we have to swallow but we still have to find ways to make it work..
>
> (Senior Design Coordinator, Design and Build Contractor, Airport megaproject).

In the quote above, the Senior Design Coordinator explained they were just following the approved design, which determined the ducting layout and led to the lack of ceiling space: "we have no space". From the contractor's perspective, they were merely adhering to the client-approved plans, which resulted in the reduced ceiling height.

They used a tactic called "rule literalness", claiming they were strictly following project rules. The contractor justified their actions by saying they were just working according to the client's approval of the initial structural plans. This approach had a ripple effect on the specialists' work. By sticking so rigidly to the rules, the contractor twisted the actual purpose of having those rules.

Box 6.8 Case study – masquerading to strategically deceive and shift responsibility.

The combination of rule literalness and the act of masquerade provided a strategic advantage, allowing the project team to shift responsibility for mistakes or failures onto others. By feigning a lack of knowledge about the situation, they could plausibly deny awareness of the necessity for their intervention. The Senior Construction Manager from the Airport megaproject client office noted that such deceptive practices were particularly prevalent among contractors.

> they [the contractor] will just pretend or ignore, they are very good at, sometimes if the issue is very straightforward they can turn it like "oh man, it's all confused" ... so they just pretend they don't know, this is not in our scope. Simple as that but no, we have to show them the contract, "look this, this, this, this". "Well my understanding is not like that", the contractor would say! Yea its (frustrating) but we have to live with this, because we've already awarded them the job.
>
> (Senior Construction Manager 2, Airport Client Office, Airport operator, Airport megaproject)

This tactic of presenting themselves as less informed than they actually were served as a protective measure, enabling them to navigate complex situations without assuming direct responsibility for certain outcomes.

Contractor's tactics to exploit time pressure

The contractors' tactics to get better outcomes were shaped by knowing the client had a very tight deadline. Knowing the importance of time, the contractor understood that the client, under tight deadlines, would probably approve any claims, changes, or proposals they made, since the project had to "proceed no matter what". With few choices left, the client had to process submissions less critically to avoid further delays.

Box 6.9 Case study – exploiting time pressure: contractor strategies in tight deadlines.

The Senior Design Coordinator from the contractor firm highlighted the risks of the project's fast-track nature. He mentioned this was the first project he'd seen go to construction based only on working drawings, instead of the usual detailed drawings. He said they had little time for decisions, and the client constantly pushed them to move forward. The coordinator explained, "we were always under pressure to keep the work going, even though the drawings weren't finalized, so we just went ahead". Despite the challenges of quick decision-making, the contractor saw an opportunity to use this pressure to their advantage. They could push the client for necessary approvals to avoid delays. The Senior Design Coordinator said, "They know the contract is not solid, but they have to proceed no matter what".

The contractor expected that due to the client's time pressure, they would likely approve any claims, changes, or proposals submitted, as they needed to keep the project moving, "no matter what". This approach of rushing the client for approvals sometimes worked in the contractor's favour. Some issues were missed by the client because there wasn't enough time for thorough decision-making. As mentioned earlier, the CDC team had approved the construction of the structural slab. The contractor had noted a problem with this, which resulted in the varied ceiling heights.

The account of the ceiling height debacle revealed the contractor's adeptness in practising destructive deviance. This was particularly evident in their skilful manoeuvring to "slip under the radar", effectively avoiding responsibility for behaviour that did not comply with standards or expectations. From the contractor's perspective, accepting the penalty for non-compliance was deemed a lesser burden compared to the complexities and costs involved in rectifying the issue to achieve compliance.

Box 6.10 Foot dragging to stall activities and redistribute power.

Foot dragging, or the strategic taking of time, represents a nuanced form of resistance, intricately balancing compliance with subtle defiance of established time rationalisations (Ewick & Silbey, 2003). This tactic involves manipulating the time allocated by the party who typically dictates or controls behaviour in temporal terms. In the case of the ceiling height debacle, this time belonged to the client, who was operating under the typically stringent deadlines characteristic of megaprojects. The contractor recognised these time constraints as opportunities to subtly disrupt and redistribute power dynamics, aiming not only to gain a measure of control over various situations but also to use it strategically to their benefit. Consequently, foot dragging emerged as a deliberate tactic by the contractor to slow down project activities. This intentional stalling was aimed at constraining the range of decision-making options available to the client, thereby altering the power balance within the project dynamic.

Permissive ethical culture resulting in normalisation of deviance

Acclimatisation to deviant behaviour occurs gradually, with the normalisation of deviance unfolding over time, often facilitated by the client's failure to challenge questionable practices. When questionable practices are progressively accepted, they eventually become entrenched in standard procedures. Clients, due to their influential position and power in projects, are typically expected to uphold and maintain the fundamental fairness and integrity of the projects. Their actions, or lack thereof, in addressing deviant practices play a critical role in shaping a project's ethical standards.

Box 6.11 Case study – resignation and acceptance of deceptive practices culminating in the normalisation of deviance.

Members of the client organisation's CDC team conveyed a resigned attitude while expressing their discontent regarding the ceiling height of the terminal building.

> Instead of 2.8m it becomes 2.4…we have to accept the situation. We have to deduct some money from the contractor because of non-compliance… So, in this case we've had to compromise, so we'll penalise the contractor.
>
> (General Manager, Head of Construction, Airport Client Office, Airport operator, Airport megaproject)

> The ceiling height is very low; you don't feel very comfortable… you feel depressed you know…People will experience that. We cannot change it now because of the time and the requirements. Nobody looked at it seriously maybe…
>
> (Senior Project Architect, Client Design Consultant, Architecture firm, Airport megaproject)

Although dissatisfied with the final result, the CDC team recognised the inevitability of the situation. They perceived that there were limited options available, other than penalising the contractor for non-compliance. However, the issue of the varying ceiling height remains a long-term concern, as it will continue to affect the use of the space for years.

The architect from the CDC team acknowledged that despite being aware of certain issues in the project, such as those related to the structural system, a more lenient approach was adopted in reviewing the contractor's work. This approach stemmed from an understanding that requesting changes to the work already completed by the contractor would lead to extended project timelines. Consequently, the client team felt constrained in their ability to make significant amendments or take alternative actions, opting instead to "follow the normal procedure but to be less critical". This change in approach did not occur abruptly but evolved over an extended period. This suggests that this gradual shift led to the normalisation of deviance within the project's operational framework over time.

> When they proposed the structural system, our structural system commented on that, but they never listened. So, what can we do? We cannot amend or do anything from there…So that's been for a long time… you can change but it takes a longer time… But actually, when you have time earlier you didn't expedite, when we asked you to bring in more manpower you didn't bring in. When we went to the site a few people only working.
>
> (Senior Project Architect, Client Design Consultant, Architecture firm, Airport megaproject)

> What are you doing for the last 1.5 years? You don't really focus, you don't really concentrate…you don't bring in manpower, or materials, everything, you just ding-dong ding dong talking about contract, oh in our contract we don't do this…they don't behave like they are completing the job so now what can we do? We can only say approve, approve.
>
> (Senior Architect, Consultant Design Consultant, Architecture firm, Airport megaproject)

The quotes in Box 6.11 illustrate the client's reluctance to exercise their authority to interrupt the gradual normalisation of deviant practices. The CDC team, as evidenced by their responses and actions, displayed a tendency to acquiesce to non-compliant behaviour that fell short of meeting the project's requirements. Statements such as "they never listened…what can we do?" and "when we asked you to bring in more manpower, you didn't bring in" reflect a sense of resignation. This stance suggests an implicit recognition of the contractor's anticipated behaviour, leading to a tacit tolerance of their deviant actions, which eventually became normalised.

The Senior Architect from the client team acknowledged the impracticability of confronting the contract, despite being cognisant of their strategic use of delaying tactics in the initial phases of the project. In the interest of progressing the project, the Senior Architect indicated that the client team was prepared to condone a certain degree of undesirable behaviour, as encapsulated in the sentiment "we can only say approve, approve". This approach highlights a critical aspect of the client-team dynamic, where the desire to advance the project overshadowed the imperative to uphold stringent compliance and ethical standards.

Chapter summary

The stories featured in this chapter emphasise that while deception can be a powerful tool in the hands of project managers, it requires a careful, ethically informed approach. The lessons drawn from these case studies are vital for future megaproject endeavours, advocating for a balanced approach that considers both the ethical implications and the practical necessities of managing megaprojects. The ultimate takeaway is the need for a heightened awareness and strategic, yet ethical, approach to decision-making in the complex, multi-faceted world of megaprojects.

References

Donzelot, J. (1979). *The policing of families*. New York: Pantheon Books.

Ewick, P., & Silbey, S. (2003). Narrating social structure: Stories of resistance to legal authority. *American Journal of Sociology*, 108(6), 1328–1372.

Hajikazemi, S., Aaltonen, K., Ahola, T., Aarseth, W., & Andersen, B. (2020). Normalising deviance in construction project organizations: A case study on the collapse of Carillion. *Construction Management and Economics*, 38(12), 1122–1138.

Winch, G. (2017). Megaproject stakeholder management. In B. Flyvbjerg (Ed.), *Oxford handbook of megaproject management* (pp. 339–361). Oxford: Oxford University Press.

7 Recklessness

Destructive deviance can be detrimental to project outcomes

Recklessness: the dangers of low-governance environments

Quadrant 3, known as recklessness, combines low governance with harmful deviant practices. While high autonomy can spark creativity, it can lead to problems when combined with self-serving, harmful actions. This is because environments with weak rules can let questionable practices grow unchecked, even becoming normal and accepted.

In such settings, people may act recklessly, focusing on their own interests without considering the damage to the project, especially if they face no serious consequences. This chapter documents a case study where the contractor deliberately ignored authority for their benefit. Their actions, meant to influence decisions for their own goals, didn't have immediate negative effects, leading to a pattern of harmful behaviour becoming standard.

Therefore, this quadrant is risky and needs careful monitoring by those in power to prevent the normalisation of such destructive behaviour. It is crucial to stop these practices from gradually becoming accepted.

Roof redesign

The following sections focus on the story behind the redesign of the terminal building's roof at an airport project in Asia. They detail the events leading to the final roof design, which was much flatter and less striking than the original concept. The original design, chosen as the winning entry in a competition, was seen by the Client Design Team as more symbolic, adding aesthetic value rather than serving a functional purpose. However, assessing architectural design often involves subjective judgments, making it hard to justify design decisions definitively. The contractor viewed the roof design as negotiable and eventually convinced the client to approve a redesign by using deviant tactics.

Box 7.1 Case study – Value engineering exercise leading to roof redesign.

Under the Design and Build contract, the contractor had the option to suggest different design solutions if they could show that these changes would

DOI: 10.1201/9781003377481-7

enhance value. The client viewed this as a strategy to enhance the megaproject's overall performance. However, for the contractor firm, the main goal was to complete the project at the lowest cost. Motivated by financial considerations, the contractor identified the roof design as an area where they could significantly reduce project expenses.

> Our requirement is first thing is monetary, economical and also technology that comes with the money. So, our specialist said that kind of design is expensive so [to] try to reduce the roof whereby the basic form is there but the geometry is different.
>
> (Senior Design Coordinator, Design and Build Contractor, Airport megaproject)

The contractor firm proposed a redesign of the roof, which was seen as a significant change from the original plan:

> the design is now different, as you can see the previous design is very dynamic, it's very like a spaceship… when it comes to the real thing the design is no longer following the original concept…And it's because of cost and time.
>
> (Senior Project Architect, Client Design Consultant, Architecture firm, Airport megaproject)

As the Senior Project Architect from the Client Design Consultant (CDC) team noted, the change was driven by cost and time considerations. The contractor's approach, focusing on value engineering, was primarily aimed at short-term financial savings, without much regard for the original design's purpose.

Cost-cutting disguised as regulatory compliance in roof redesign

The contractor, who led the value engineering process, explained that their proposal to redesign the roof of the terminal building aimed to meet several goals. They justified the need to change the original roof design by citing a requirement from the regulatory body: ensuring the control tower had the proper line-of-sight to the aircraft. Thus, their argument was based on a technical reason, but in reality, the main focus was on cutting costs.

Box 7.2 Lesson – Technical reasoning can be used to mask hidden agenda.

When discussing the main motive for the roof redesign, the Senior Design Manager from the airport project client appeared convinced that the redesign was in response to regulatory requirements:

> So now the roof looks a little bit flat. It's just different. So, let's say you want to design your hair to be the afro look then suddenly you have flat hair so something like that. [Interviewer: So, it [the roof redesign] wasn't a cost issue?] No, it's mainly the constraints – the ICAO [International Civil Aviation Organisation] requirements in terms of the design…due to the line of sight where the control tower needs to see the aircraft, which is the requirement of the control tower, you must see the width on the taxiway.
>
> (Design Manager and QAQC Manager, Airport client office, Airport operator, Airport megaproject)

To the Design Manager, altering the roof design was viewed as a minor modification, akin to changing a hairstyle. However, the CDC team had a different viewpoint on this change.

> The design is developed by the contractor and of course they know what's the best way to cut down the cost, so they used the guidelines, the restrictions from DCA [Department of Civil Aviation] to get the cost cutting exercise approved. This could be a game from the contractor side…there are ways to maintain the shape, but…the contractor managed to convince the client. So, it's become a major item for the cost cutting exercise.
>
> (Resident architect, Client Design Consultant, Architecture firm, Airport megaproject)

> … the design is now different, as you can see the previous design is very dynamic, it's very like a spaceship… when it comes to the real thing the design is no longer following the original concept… We won this project because it was a competition, and suddenly just in front of us is a box kind of building sheltering a commercial area which kill the aesthetics of our airport…And it's because of cost and time.
>
> (Senior Architect, Client Design Consultant, Architecture firm, Airport megaproject)

The architects from the CDC team believed that the real reason behind the roof's redesign was the contractor's aim to reduce its size, thereby saving time and costs. Members of the CDC architecture firm suggested that there were alternative ways to preserve the original, more dynamic design without just flattening it. However, the contractor was more successful in persuading the client to accept their redesign, which was primarily driven by a hidden agenda to cut project costs, ultimately compromising the airport's aesthetic appeal.

Navigating hierarchies for efficiency

In the stories shared, project team members showed an understanding of their social positions, roles, and ability to impact project results. People without sufficient authority often found ways to exert influence in situations where they appeared to have limited control.

There was an understanding that people lower in the hierarchy often lack the power to initiate actions. For project team members, instructing the "top boss" and going "straight to their boss" was seen as a much more effective strategy, compared to dealing directly with those at the lower level in organisations. Conversely, those at the lower level of organisations understood their limited power and utilised the same strategy to disrupt hierarchy to achieve their desired ends. For these players at the lower level, the game was no longer open to everyone, and thus they needed to access the arena through alternative means. By escalating issues to higher levels in organisations, project team members were able to increase the likelihood of having their needs met in a timely manner when superiors were informed.

Box 7.3 Case study – Strategically bending rules for effective project outcomes.

Project team members shared instances of stakeholders deliberately overlooking established hierarchy, authority, and their responsibilities. The General Manager of the construction team for the airport project knew that sometimes by passing the usual chain of command was necessary to get better results.

> Yes, I'll go straight to their boss (contractor). You cannot deal with the people at the lower level because these people are limited in their decision making...So I talk to the top boss. And they can give instructions after that…it's the same [the other way around], they come to me and say, "boss I have limited power, I need this to be done fast. Can you talk to my boss?"
> (General Manager – Construction, Airport client office, Airport operator, Airport megaproject)

Even though fully aware of the formal protocols and procedures, the Senior Design Coordinator from the contractor firm preferred to deal directly with the client when trying to resolve project issues:

> Let's say we found a loophole we discuss how to solve the problem without escalating the costs. We can both agree on how to solve the problem, minimal costs, practical and fulfil the basic requirements. I talk to different people…sometimes I will talk to him [client-side project manager] and ask for advice and highlight the issues…I deal directly with the client.
> (Senior Design Coordinator, Contractor firm, Airport megaproject)

Bypassing the chain of command for roof redesign approval

The contractor's skill in navigating the bureaucratic system was evident in the stories he shared, showing how easily he adapted different tactics for different situations. By using a tactic to disrupt the hierarchy and bypass official lines of authority, the contractor could directly negotiate with the client. This approach allowed for informal agreements, avoiding the formal channels that were expected to involve stricter procedures and requirements. However, there was a noticeable disregard for how such deviant behaviour might negatively impact the project.

Box 7.4 Lesson – Misguided mastery of bureaucratic navigation can compromise project integrity.

Regarding the approval process of the new "flattened" roof design, the Resident Architect from the CDC team mentioned that the contractor clearly chose to break the rules and bypass hierarchy to get the design approved.

> This has happened because sometimes contractors ask directly from people from the airport operator...actually we are the CDC…this is the frontier for the airport operator. They're the design and build contractor…So they should come to us for approvals and then we go to airport operator.
>
> (Resident architect, Client Design Consultant, Architecture firm, Airport megaproject)

Knowing that the CDC did not agree with the proposed flattened roof design, the contractor decided to bypass them and directly present their case to the client. The Resident Architect from the CDC team expressed frustration over the contractor's strategy to disrupt the hierarchy and circumvent his team, which was supposed to be the primary interface for the client. Additionally, the tight schedule of the project seemed to benefit the contractor, as finding ways to keep the original design while reducing costs would have required more time.

Urge to achieve financial goals

The contractor's strong focus on financial gain overshadowed their commitment to the project's objectives and ethical standards. The project team members noted that contractors often willingly take risks to win bids in a competitive market. This was evident in the case study airport project, where the contractor intentionally found contract loopholes to increase their earnings.

Box 7.5 Case study – Contractor tactics: initial underquoting and subsequent compensatory strategies.

Representatives from the client organisation described how the contractor initially underquoted costs to secure the project, later seeking ways to recover funds through various means.

> … they tendered at lower cost where they [now] cannot handle [the costs] anymore so that's the problem… it's just the way it is.
>
> (Senior Construction Manager, airport client office, Airport operator, Airport megaproject)

> Because of certain reasons maybe commercial, so the contractor says, "hey look, we're not making money on this project, so we need to do something about it". So even a small thing they make it an issue. They just make use of our document, not to say the document is bad, no. But as I said they will look into the grey area, where there's a grey area it's a point for them to raise an issue and try to avoid to do the scope of works. So, they make it very hard for us.
>
> (ID works manager, airport client office, Airport operator, Airport megaproject)

The quotes emphasise the financial difficulties faced by the contractor as the project advanced. Statements like "they can't handle it" and "we're not making money on this project, so we need to do something about it" were common. This situation led the contractor to focus on both cutting costs and finding ways to increase revenue. They often tried to do this by "avoiding the scope of works" or "raising issues". The client representatives frequently mentioned "contractor games", referring to various deviant strategies or tricks used by the contractor to meet their goals:

> So as a contractor they will take any opportunity to get additional time, money, additional anything.
>
> (Senior Project Architect, Client Design Consultant, Architecture firm, Airport megaproject)

> The contractor will look at different areas or avenues to collect more money… by maybe invoking variation orders, things which is not inthe original value… it's a matter of legals and legals looking at each other. The contractors are finding loopholes to go about it.
>
> (Master planner, airport client office, Airport operator, Airport megaproject)

It was evident that the contractor prioritised economic concerns above all else, likely due to the risk to the organisation's survival. Consequently, the contractor exhibited a significant level of deviant behaviour, driven by the need to meet financial targets, while showing little concern for the overall performance of the project.

Clever gamesmanship to facilitate destructive deviance

In the airport project, which was a Design and Built contract, it was understood that there would be some uncertainty in the design process and final outcome of the project. Project team members noted that the contractor had considerable leeway in interpreting the client's needs into a physical structure, as the initial concepts were generally broad and open to interpretation, allowing room for alternative designs.

Notably, the senior design coordinator from the contractor firm openly discussed his role in exploiting the highly regulated project environment. He acknowledged his tactics in identifying and utilising loopholes in the contract for financial gain:

> They call me the bad boy. I'm the one who creates problems and highlights issues in the contract… There are many loopholes in this contract. That's my goal, to find those loopholes and that's why we always disagree.
>
> (Senior Design Coordinator, Design and Build Contractor, Airport megaproject)

The Senior Design Coordinator from the contractor firm clearly understood his role as the "bad boy" and a skilled navigator of bureaucratic processes. He viewed the complex web of meetings, paperwork, and the matrix communication system not as hindrances, but as opportunities for strategic manoeuvring. The General Manager, head of Construction at the airport project client office, noted that "grey areas" in the project were often exploited by the contractor as chances to twist situations to their benefit. An example is provided in Box 7.6.

Box 7.6 Case study – Exploiting ambiguity to advance self-interest.

Understanding the power of documentary evidence, the Senior Design Coordinator explained that he strategically used the original set of drawings, which were the official tender documents, to safeguard against potential future issues.

> When they [the client] issue the Design Order (DO) that means there is changes to the design…To us, as a contractor, any changes to the drawing is claimable to us…so whatever is in that area in that drawing we already outlined they have to come back to us and say, "OK no, this is not included" and then we'll still argue, "yes we included this".
>
> (Senior Design Coordinator, Design & Build Contractor, Airport megaproject)

It was widely recognised that a key skill for navigating the bureaucratic maze effectively was to keep detailed documentary evidence. The contractor saw numerous opportunities to charge the client additional fees, relying on a precise record of prior agreements and understanding that "any changes to the drawing were claimable". Thus, meticulous record-keeping of Design orders from the client was crucial. This allowed them to show that alterations were made to the original design, justifying extra costs. In this manner, the contractor cleverly leveraged formal documentation to their advantage. When queried about the difficulty of getting his claims approved, the contractor's response was:

> No. The thing is you must substantiate all those claims. They will evaluate and the other person, as the QS, they want it in detail and if it's all in detail and they can evaluate then they will forward their recommendation, it's easy… For us whatever we build we must get paid.
>
> (Senior Design Coordinator, Design and Build Contractor, Airport megaproject)

The Senior Design Coordinator's response to whether it was hard to get approval for his claims – simply "no" and "its easy" – highlights his expertise and experience in skilfully bending the rules. He described the quantity surveyor's evaluation process, emphasising the need for detailed claims, in a direct and matter-of-fact way. This approach, which he understood as an "easy" method to reach his objectives, reflects his implicit knowledge of how to navigate the system effectively.

The client's role in establishing an ethical framework in megaproject decision-making environments

The client's responsibility in megaproject extends to shaping an ethical decision-making environment, setting the standards, and guiding principles for the project. Their role is crucial in defining and maintaining the ethical framework that governs project decisions and actions.

Box 7.7 Importance of client in setting anti-permissive ethical climate.

The client's representatives acknowledged the contractor's ability to successfully secure approvals for their claims.

> So, they will just try to give us as many VOs as they can. So, let's say they send us 1 million and maybe they can get 200,000… I think it's any project but this [project] is a great opportunity to do it.
>
> (Resident Architect, Client Design Consultant, Architecture firm, Airport megaproject)

> On any project, contractors have strategies to get any additional VO [variation orders] …as consultants we have to face these issues and manage them properly and try to reduce as much as we can… we have to understand the contracts and know how to use contracts.
>
> (Senior Project architect, Client Design Consultant, Architecture firm, Airport megaproject)

> Talking about the behaviour of contractor, so a contractor will always be contractor, they have their own ways. They will always be like that and sometimes it's difficult and sometimes things don't go the way we want so we just yell at them (laughs!) It requires yelling, letter writing and things like that.
>
> (Senior Construction Manager, airport client office, Airport operator, Airport megaproject)

The client representatives' understanding of contractual terms, and critically, their response to the contractor's manipulation of the contract, were key to managing the high number of claims made by the contractor. The Senior Construction Manager from the airport project client office emphasised the importance of the client being adept at using different strategies to control the contractor's behaviour, which included assertive actions like "yelling, letter writing and things like that".

Chapter summary

The contractor's adeptness in destructive deviance was evident in their ability to persuade the client to approve a roof redesign, despite the disapproval of the CDC team. The contractor strategically formulated a compelling argument for the redesign, knowing precisely what type of information or reasoning would sway the client in their favour. While the CDC architecture firm suggested alternatives to preserve the original roof's dynamic design, they were ultimately overshadowed by the contractor's persuasive strategy. Moreover, the contractor's skill in taking independent strategic actions extended beyond interactions with clients.

The roof redesign story presented in this chapter highlights that even well-intentioned project team members can be at a disadvantage if they don't present the right argument. The contractor's key tactic was framing the redesign as necessary for regulatory compliance. This approach was more effective than an aesthetic argument, as it was easier to justify a "need" rather than a "want". There are often stereotypes of about professions, including the belief that architects favour aesthetics over functionality and cost. This stereotype was evident in the roof redesign saga, and the contractor skilfully used this perception to their advantage.

8 Entrepreneurial

Authentic use of power can enhance project performance

Entrepreneurial

Quadrant 1, also known as "entrepreneurial", fosters an ideal environment for megaproject decision-making. Here, constructive deviance is encouraged while there is a more flexible approach to enforcing rules, regulations, or systems. The entrepreneurial quadrant advocates for occasionally bending standard rules or procedures, not for the sake of rebellion, but as a means to foster innovation and problem-solving.

Constructive deviance, in this context, is critical in enhancing creativity, thereby aiding in tackling challenges and improving the overall performance of megaprojects. In practice, this could involve project team members adapting or working around established processes – such as controlling information flow or challenging traditional hierarchical structures – to better suit the unique demands of the project. These deviations, though they might appear to contravene established protocols, are often instrumental in ensuring the smooth progress and timely completion of projects. They can lead to innovative solutions unattainable through conventional methods.

What is important to note is that these acts of deviance are not acts of defiance or dishonesty. Rather, they represent a proactive approach by project team members, who take the initiative to navigate through and surmount project challenges.

The idea that breaking rules can sometimes be beneficial is supported by academic research (Spreitzer & Sonenshein, 2004; Bodankin & Tziner, 2009). It suggests that despite seeming to violate prescribed rules, when undertaken judiciously, such behaviour is often functional and significantly contributes to achieving the objectives of a megaproject.

When rule-breaking serves a greater good

Building upon the story introduced in Chapter 5, this section details the latter stages of the journey towards regulatory approval for the bowl cooling system. The project team encountered a significant obstacle in obtaining regulatory approval for the innovative bowl cooling system. This challenge did not stem from any shortcomings in the technical soundness of their design. Instead, it was the result of wider political factors that extended beyond the scope of their technical expertise.

DOI: 10.1201/9781003377481-8

The path to achieving approval highlights a critical insight: strict adherence to rules is sometimes not sufficient to achieve the desired outcome.

The story unfolds to reveal that the project team found it necessary to deviate from established guidelines. This departure was not a choice but a requirement, highlighting the limitations of strictly rule-based approaches in addressing complex issues, as shown in Boxes 8.1 and 8.2.

Box 8.1 Case study – Knowing when to break the rules.

The project team felt that the game was no longer open to everyone; the ability to make a difference was now only in the hands of those with authority. This is shown in their use of deviant tactics during "bottleneck" situations that were "out of their control". They viewed these alternative approaches or tactics as necessary to solve immediate problems, thereby minimising delays and financial losses. The project team acknowledged their limitations in negotiating with the relevant regulatory body:

> When we've come into a bottleneck then that's qualified as maybe a political decision you know and it's out of our control so then we go to the government agency and say, "you government to government, please talk". This was one of the things where the government agency has to be involved so that everybody was reasonable but just to make sure otherwise you could be in a situation where the contractor in front of the authority who do you think will win? [Laughs] So they see how we [they] can do sort of [be the] mediator. But our attitude is we shouldn't abuse such things
>
> (Senior Technical Director, Design and Build contractor, Sports facility megaproject).

The power dynamics were clearly on display here, particularly the influence afforded to the government agency client simply because it is a government agency. This allowed the client to engage in direct discussions with the regulatory body, effectively exercising a form of "invisible authority". The Senior Technical Director humorously acknowledged this power imbalance when he quipped about putting a contractor and an authority figure head-to-head to see "who will win".

Addressing the innovative aspects of the bowl cooling system, the Senior Manager of the Sports Facilities division advocated for a more flexible approach from local authorities. Rather than sticking rigidly to existing guidelines, they suggested that the system should be evaluated according to negotiated criteria tailored to its unique features.

> It's the first time that we're going to cool a sports facility this way in this country, whether it is the building authority or the fire authority,

> they had their concerns…on this basis we had to work out a lot of modelling but of course if they followed the local guides…they would fail to comply… when…you know they [design team] need some help… we'll go behind them and talk to the agency like in this case… that's why we were able to convince them to say why we need this…But of course we had to pre-agree with them…on how we can operate it and things like that
>
> (Senior Manager and Head of Design and Technical – Sports Facilities Division, Sports government agency, Sports facility megaproject)

In the end, it was the Sports government agency's private negotiations with the regulatory body that secured approval for the bowl cooling system. This occurred despite the design team's substantial effort to justify the system based on technical know-how. The regulatory body seemed to give more weight to the sports government agency's position and arguments.

The behaviour of both the client and project team members was intentional and went against official protocols for communication. However, these actions were taken with the belief that they would ultimately benefit the project as a whole.

Box 8.2 Lesson – Breaking the rules is sometimes needed to make things work.

Given the strict and bureaucratic nature of the regulatory body, the project team recognised the need to circumvent usual channels. They understood the need to take the bowl cooling issue "up to a higher level" for resolution.

> The system is so regimented...the way that the regulatory body is very strictly doing something and when you do something different people don't know how to deal with it. So, it means you'll have to bring it up to a higher level...So that level of thinking is thinking about how can we reach a compromise … and that compromise can only be discussed at that high level
>
> (Senior Project and Design Manager, Architecture firm, Sports facility megaproject)

> At one point we didn't have clarity, that we were in violation of regulation as the regulatory body was interpreting it and the sports government agency was working behind the scenes to work through that…they do kind of help go to government departments and help smooth things out.
>
> (Senior Design Manager, Design and Build Contractor, Sports facility megaproject)

> they (Sports government agency) provide assistance...there are discussions going on between quite senior people within the Sports government agency...the regulatory body...ministry level for the Ministry of Resources, Ministry of Environment... there have been discussions going on between those sort of groups at quite high level
>
> (Senior Associate, Sports Architecture Design Firm, Sports facility megaproject)

The Senior Manager from the Sports government agency's Sports Facilities division, who was responsible for design and technical matters, confirmed their involvement in informal discussions with the regulatory body. These conversations aimed to secure approval for the bowl cooling system. The Senior Manager emphasised the need to "to the top", by passing usual bureaucratic channels to achieve the necessary flexibility. Being a government agency themselves, they had the unique advantage of directly negotiating with senior figures in the regulatory body to find a compromise.

> So, it's a matter of making big decisions at a different level sometimes. To resolve things...higher level whenever it's necessary...as long as we think that they have hit a stumble block then we have to step in. So, we really have to go with them either together with them or from behind
>
> (Senior Manager and Head of Design and Technical – Sports Facilities division, Sports government agency, Sports facility megaproject).

After agreeing to the renewable energy condition, the design team and their partners faced a new challenge: how to finance and implement the photovoltaic (PV) cells needed to offset the bowl cooling system's energy consumption. This was a curveball that was not budgeted for.

Reaching compromise solutions

In the context of large-scale megaprojects, regulatory agencies typically have the authority to interpret how rules apply. Specifically, the regulatory body imposed additional requirements for the high-level awards of GreenMark GoldPlus and Platinum, elevating the standards the Sports facility megaproject had to meet to achieve such recognition. During the initial review, the regulatory body suggested a trade-off: to win the GoldPlus award, the design team had to ensure that the energy for the bowl cooling system came entirely from renewable sources. This allowed the regulatory body to assure the public that the cooling system was environmentally friendly. For the design team, this was a fair exchange that helped the regulatory body to manage public perception. The Senior Design Manager put it as a "straight swap" where both parties found a mutually beneficial solution to a complex issue.

Box 8.3 Case study – A balancing act: navigating bureaucracy through strategic compromises.

The Senior Project and Design Manager from the architecture firm highlighted a key strategy of the design team. He noted that the team was conscious of the need to make concessions, but they were careful not to concede more than was absolutely necessary. He explained, "they knew they had to give something, but they didn't want to give more than what was essential".

> One thing is trying to see things from a regulatory body's point of view, what are the things that concerns them… How do we make it that their concerns are mitigated in some fashion…I think that's what AAA [the architecture firm] brought which was…understanding what concerns are about…is to talk the regulatory body through how to view the issue technically…to manage public awareness as well because in the end government bodies, that being their main concern… that's the way in which AAA has managed to help the regulatory body connect all the dots.
>
> (Senior Project and Design Manager, Architecture firm, Sports facility megaproject)

The Senior Project and Design Manager emphasised the importance of understanding both the regulatory body's specific concerns about the bowl cooling system and the broader context of these concerns. A significant contribution of the local architecture firm to the project was their in-depth knowledge of the bureaucratic workings of government authorities. They recognised that government officials are primarily focused on adhering to rules to avoid public criticism and to main their image of propriety in the public eye.

Understanding the limitations posed by bureaucratic processes, the project team adapted their approach. Rather than attempting to change the bureaucratic system, they learned to work within it for the project's advantage. The Senior Project and Design Manager pointed out that their firm's ability to pinpoint the regulatory body's concerns and to develop corresponding "mitigation measures" was critical. This strategy enabled the project team to assist the regulatory body in "connecting the dots" needed for regulatory approval of the bowl cooling system. In doing so, they ensured that the regulatory body could grant approval while also maintaining its public image of doing the right thing.

Chapter summary

Navigating the decision-making environment in megaprojects requires a nuanced understanding of rules, the flexibility to adapt strategies, and the ability to engage in effective negotiations. The "entrepreneurial" quadrant in the governmentality framework for megaproject decision-making champions the idea of bending rules not for rebellion but as a catalyst for creativity and better project outcomes. Through the use of case studies, this chapter demonstrated how such deviations from norms, though appearing counterintuitive, can lead to innovative solutions and enhance project performance.

References

Bodankin, M., & Tziner, A. (2009). Constructive deviance, destructive deviance and personality: How do they interrelate? *Amfiteatru Economic*, 11(26), 549–564.

Spreitzer, G.M., & Sonenshein, S. (2004). Toward the construct definition of positive deviance. *American Behavioral Scientist*, 47(6), 828–847.

9 Governmentality as an alternative approach to decision-making

How to support constructive deviance

Documenting authority: using textualisation for project governance and decision-making

Setting the correct ethical climate on a megaproject is of paramount importance to facilitate sound decision-making. This guiding foundation serves as a guiding principle to promote responsible and ethical actions throughout the project's lifecycle. It not only helps maintain integrity but also fosters trust among stakeholders, ultimately contributing to the project's success.

Rules and regulations serve to delineate the boundaries within which project team members must operate. While these constraints can seem restrictive, they are instrumental in "setting the tone" and shaping the decision-making process in projects. The reliance on formal governance mechanisms for behaviour control and compliance enforcement is heavily anchored in the concept of textualisation. Project team members acknowledge its pivotal role as the core organising factor in projects. For many project team members, the authority of written documentation takes precedence over verbal accounts. This preference aligns with previous research by Ewick and Silbey (1998), which suggests that the greater authority of written texts stems from their ability to be preserved, retrieved, and scrutinised.

Box 9.1 Lesson – Establish clear rules and regulations to foster trust and predictability in megaproject teams.

The impartial nature of written policies, procedures, and record-keeping protocols are viewed as the key guiding principles for determining appropriate actions and behaviours of project team members. Megaproject team members, bound by the project contract, operate within defined constraints and adhere to established rules and regulations, instils trust in a fair and impartial system to govern project behaviour. This trust is evident in various statements from team members from two megaprojects such as:

> "they are not supposed to just do anything"
> "this is how we work"

DOI: 10.1201/9781003377481-9

"what meetings I go to or don't go to, its all defined"
"meetings and paperwork…we have to go through all the processes"
"there is a rule"
"it's all clearly spelt out"
"they should come to us for approvals"
"they have to follow our contract"
"we have to follow the book"

These reflections indicate the importance of project team members viewing existing rules and regulations as intractable facts. They perceive these rules as impartial and effective tools for fostering predictable behaviour and curbing deviance. It is, therefore, critical to establish clear rules and regulations in the form of written policies, procedures, and record-keeping protocols. By doing so, a framework of transparency and accountability is created, guiding project team members towards consistent, ethical, and responsible actions.

Clients and regulators' roles in upholding standards

Both clients and regulators can play a crucial role in influencing the level and type of deviance undertaken on megaprojects through the ethical climate they set on projects. While the need to satisfy the strict requirements of various bodies can be very time consuming for project team members, it can also be particularly important for setting an anti-permissive climate in terms of what is deemed unacceptable behaviour in the projects.

Clients establish a unique culture within which project team members operate, by defining the parameters for decision-making related to budgets, design, and procurement. The values that clients ascribe to their daily practices inevitably condition how they act economically, subsequently affecting decision-making in megaprojects. The manner in which these values and associated expected behaviours are communicated to project team members can greatly influence their effectiveness.

Box 9.2 Lesson – Upholding standards: the crucial role of enforcing rules and procedures in megaprojects.

It is essential to not only introduce formal governance mechanisms but also enforce them effectively. Any behaviour deviating from established rules must be considered unacceptable. This principle was exemplified by the client in the airport megaproject case study, whereby individuals failing to adhere to prescribed procedures were consequently removed from the project. This action highlights the importance of compliance and the serious consequences of non-conformity in megaproject environments.

> This project is famous for the removal of people. When they are not following the procedures, they have to be removed. Two project

> managers and one project director have been removed from the JV. [Interviewer: What did they do?] Acting incompetent, with decisions, or simply they are not pleased with the answers – cos they are mumbling during meeting, indecision, no commitment, you can answer it straight, but you are always showing dissatisfaction.
>
> (Senior Design Coordinator, Design and Build Contractor, airport megaproject)

The enforcement of rules through widely publicised channels can significantly bolster the client's power and authority. In this case, the client effectively conveyed a strong message to the project team by demonstrating a zero-tolerance policy towards non-compliance. By taking decisive action – "when they are not following the procedures they have to be removed" – the client reinforced the importance of adhering to established rules and procedures, thereby ensuring a disciplined and compliant project environment.

Nevertheless, the inherently open-ended nature of project governance structures can sometimes be exploited for self-interest. This openness also creates avenues for opportunistic behaviour and harmful deviance. An example is observed in the airport megaproject, where the contractor's Machiavellianism tendencies were evident. They exhibited the cunning behaviour of a seasoned strategist, skilfully bending the rules to their advantage and strategically orchestrating actions for personal benefit. In such contexts, regulators are pivotal in creating a stringent ethical environment that serves to disrupt and deter harmful deviant behaviours. Their role is crucial in enforcing standards and practices that uphold integrity and ethical conduct within the project framework.

Chapter 7 demonstrates that if clients and regulators do not actively use their authority to interrupt the normalisation of deviant practices, they inadvertently contribute to the normalisation of counterproductive behaviours. Such inaction can pave the way for detrimental outcomes in megaprojects, highlighting the critical role of clients and regulators in maintaining ethical standards and productive practices.

Local authorities and agencies, though not formally part of project processes, become integrally involved, thereby gaining a certain level of influence over the extent and nature of deviant behaviour within megaprojects. A variety of participants, often indirect, informal, or sometimes lacking specific expertise, assume roles akin to "governmental authorities" in megaprojects. They significantly influence project directions through their requirements and play a pivotal role in establishing the ethical climate. Megaprojects, due to their scale and impact, attract considerable attention from various government agencies. These regulatory bodies, with their expertise and regulatory power, often take on critical governing roles, representing public interests. Importantly, regulators also have the responsibility of setting the ethical tone for projects, and establishing the foundational norms and values that guide behaviour and decision-making.

Box 9.3 Case study – Regulatory disparities and ethical climates: a comparative analysis of the Sports facility and airport megaprojects.

The influence of regulators in establishing a strict ethical climate was a recurring theme among project team members from the Sports facility megaproject. In contrast, only a single project team member from the airport megaproject reported challenges due to stringent requirements imposed by government authorities. The rules and regulations, as experienced by the Sports facility megaproject, created a highly restrictive environment, offering limited flexibility in advancing project matters.

> The authorities of this country have got very strong power. We got to meet their expectations first. Everybody needs to compromise a bit but usually the authorities are difficult to compromise.
>
> (Technical Director – Design and Build Contractor, Sports facility megaproject)

In certain contexts, rules may offer opportunities for resistance, such as identifying loopholes within a network of regulations. However, this scenario was not applicable in the highly rigid and inflexible context of the governing authorities. The regulators demonstrated a low tolerance for deviations from their standards.

> the difficult thing with the regulatory body is because there's such a repetitive nature to the way that things are done here in this country… also in defence of the way that the regulatory body does things I think because developers here are so aggressive it forces the regulatory body to say this is the line you follow. So, because the regulatory body then says this is the line you follow, all of the regulatory body's staff that we deal with on a day-to-day basis are very well trained in the things we have to do.
>
> (Senior Project and Design Manager, Architecture firm, Sports facility megaproject)

> We're going through just demonstrating compliance for the things which the regulatory body have queried when we went through the the assessment…If we have to spend a little bit of energy in doing that, we have to spend a little bit of energy in doing that…"
>
> (Senior Associate, Sports Architecture firm, Sports facility megaproject)

Project team members highlighted the impracticality of contesting the regulatory body, suggesting it was more feasible to comply with the established "line to follow". They acknowledged compliance as an unavoidable aspect of their work, accepting that it requires a certain amount of effort. Adding a comparative perspective, the Chief Operations Officer from the airport megaproject noted the relative strength of regulators in other countries compared to their counterparts in his country, indicating a difference in the regulatory power exerted in different countries.

> So, you can see the difference between other countries and ours. That country is more dominant. You do my way, there's no alternative. Here, it's different because probably the government's a little lenient.
>
> (Chief Operations Officer, Airport operator, Airport megaproject).

This situation hinted at a stark difference in regulatory approaches: authorities in the sports facility megaproject exhibited little tolerance for deviations from rules, whereas authorities in the airport megaproject appeared more lenient towards questionable practices. Such disparities in regulatory enforcement significantly influenced the ethical climate within the decision-making processes of these megaprojects.

Balancing act: navigating the complexities of deviance and autonomy in megaproject decision-making

Despite the presence of highly structured project contracts, protocols, and procedures, these tools are insufficient in tackling the complexities and challenges encountered in megaprojects. The governance documents, written with a level of generality to avoid being overly prescriptive, inherently suffer from a lack of clarity. They are riddled with contradictions and tensions, as evidenced by the experiences detailed in the case studies in previous chapters. This ambiguity in the written guidelines opens up opportunities for project team members to employ a variety of deviant tactics as they seek to resolve project issues. This situation highlights the limitations of formal governance structures in addressing the intricate realities of megaproject decision-making.

Further to this, megaproject team members are not merely passive recipients of regulation; they also play active roles as agents with rights and the ability to influence project outcomes using various deviant tactics. As shown in the case studies governing is undertaken *through* these everyday agents rather than simply *on* them. The formal contracts and procedures governing megaprojects, with their hierarchical power structure, might suggest that team members lack the freedom to act independently and autonomously. However, contrary to being seen as mere objects of regulation, bound to their roles in a mechanical "thing-like" manner, these individuals demonstrate agency and influence within the project's framework, subtly shaping its course through their actions and decisions.

The role that autonomy plays in bringing about both constructive and destructive deviance is important to consider in this context. In the two case study megaprojects, the information and rules governing decision-making were open to interpretation, selective application, and strategic (mis)representation. Within this framework, project team members were afforded a considerable degree of autonomy and freedom. This latitude enabled them not only to adhere to the rules but also to manipulate them to their advantage. The stories shared by megaproject team members illustrated their involvement in a game-like environment of contestation, where the outcomes were heavily influenced by the skills, experiences, and knowledge of the individuals involved. Consequently, this environment fostered a perception of some players being more adept than others, leading to a dynamic of winners and losers in various scenarios.

Some facilitating factors significantly influence the extent and nature of deviance in megaprojects. The advancement of specific interests is often enabled by employing various deviant tactics, which are supported by these underlying factors. Given this context, it becomes crucial to understand the dynamics of who can do what to either foster constructive deviance or disrupt destructive deviance. This knowledge is key in ensuring while innovative and beneficial deviations are encouraged, harmful and unethical practices are identified and mitigated. Understanding these factors and their influence on project outcomes is essential for achieving more favourable megaproject outcomes.

Using covert power to navigate complex megaproject decision-making

Project team members' ability to navigate the decision-making environment in megaprojects heavily relies on the silent employment of deviant tactics. These tactics are often unspoken and taken for granted as part of the normative practices within these complex project settings.

Similar to Garland's (1997) observations in crime control, the decision-making behaviour of megaproject team members is more practical than theoretical, focused on making things work. Project team members recognise the necessity of wielding a subtle form of power, akin to what Foucault (1980) described as "less legal in character". Consequently, these practices, viewed as quasi-legal, are intentionally hidden, involving meticulous planning and execution to stay unrecognised and undetected. This approach reflects a strategic, yet understated, exercise of power, essential for navigating the complex dynamics of megaproject environments.

Box 9.4 Lesson – Unseen influence: understanding the role of covert power in megaproject decision-making.

The use of deviant tactics in megaprojects is often subtle and covert, making it challenging to detect these practices. Project team members typically employ these tactics quietly, allowing them to operate "under the radar".

This discreet approach helps them avoid detection and scrutiny, further complicating the task of identifying and addressing deviant behaviour.

> So sometimes you can't put everything absolutely on the table upfront you have to put it out bit by bit
>
> (Senior Design Manager, Design and Build contractor, Sports facility megaproject)

> So, if there's something really at that level, they try to help us behind the scenes
>
> (Senior Design Manager, Design and Build contractor, Sports facility megaproject)

> Some conversations are held quietly to try and prepare to get someone else ready to be bringing the information out in the form which is going to work for everybody
>
> (Senior Design Manager, Design and Build contractor, Sports facility megaproject)

> Let's say we found a loophole we discuss how to solve the problem without escalating the costs. We can both agree on how to solve the problem, minimal costs, practical and fulfil the basic requirements.
>
> (Senior Design Coordinator, Design and Build Contractor, Airport megaproject)

These quotes illustrating everyday deviance in the context of megaproject decision-making are fundamentally about power dynamics. They show how deviant tactics, characterised by their silent and often unnoticed execution, play a significant role in shaping decisions. This subtle form of power is described in the literature as "hidden structures of power", "deep structures", or "ways of operating". These terms reflect how such practices are tacitly accepted and thus perpetuated within the industry. Researchers like Ninan et al. (2019), Clegg (1975), and de Certeau (1984) have identified these understated yet influential practices. The routine use of these deviant tactics in everyday decision-making processes is driven by the exercise of covert power, highlighting the complex and often hidden nature of power dynamics in megaprojects.

In their practice of either constructive or destructive deviance, participants from both case studies explain their decisions to silently solicit agreements through "backdoor conversations", or to inventively use formal documents to substantiate payment claims.

The following sections offer valuable insights by showcasing strategies and tactics that can effectively intervene in destructive deviance or promote constructive deviance. These lessons provide practical approaches for addressing deviant behaviour in a manner that encourages positive and productive outcomes.

Gaining access to legitimacy and power

Megaproject team members draw from a common pool of sociocultural resources encompassing organisational, disciplinary, symbolic, and professional elements. Certain stakeholders, owing to their advantageous positions, have enhanced access to these resources. While power is not a tangible asset one can possess, individuals in certain social positions are more likely to realise their goals (Ewick & Silbey, 2003; Foucault, 1980).

Legitimacy, as a form of power, stems from formal positions within an organisation or network. This status endows individuals with inherent rights to access resources, organise processes, and exert influence over others (Yukl & Tracey, 1992). In megaprojects, legitimacy is a crucial power form, often manifested through a client's role within the regulatory framework, granting them significant sway and authority.

The central role of clients within project networks has been well-documented (Tzortzopoulos et al., 2006; Boyd & Chinyio, 2006; Crawford et al., 2008; Haugbolle & Boyd, 2016). In the fast-paced decision-making environment of megaprojects, these clients, often government agencies, possess the ability to expedite processes, leveraging their network position to grant legitimacy. Government authorities typically operate under the assumption that regulations and procedures are neutral and objective. This view posits that rules are set and enforced independently of any personal interests or positions of individuals or organisations. As Ewick and Silbey (2003) observe, this perceived impartiality is often seen as the foundation of these authorities' legitimacy and power.

Box 9.5 Lesson – Leveraging authority through hierarchical influence and strategic positioning.

The positions individuals hold within the hierarchy of an organisational network can be strategically utilised to achieve their objectives. Those who have access to a wealth of knowledge and resources are generally in a more advantageous position to "play the game" effectively. This concept is especially evident in the context of authoritative or positional power held by clients. Individuals in such roles wield significant influence, enabling them to direct project outcomes more effectively than those lower in the hierarchy. This recognition of the client's positional power highlights the importance of hierarchy in organisational dynamics and decision-making processes.

> DCA [Civil Aviation Authority), they look to the client as an airport operator. So whatever the airport operator says they (DCA) trust…so

> that's why the consultant, when they talk to DCA they always ask us to come along…so the consultant sits there and sometimes they advise technically but most of the time they are the one doing the documentation, the paperwork, whatever the airport design standard that's required…then after that the consultant will follow up with their documentation, with their submission to get it approved.
>
> (Design Manager and Quality Assurance & Quality Control Manager, airport client office, Airport operator, Airport megaproject)

> We need some support, so "Sports government agency…can you sit with us and we see how we can progress on that?" When we've come into a bottleneck then that's qualified as maybe a political decision you know and it's out of our control so then we go to the sports government agency and say, "you government to government, please talk". This was one of the things where the sports government agency has to be involved so that everybody was reasonable but just to make sure otherwise you could be in a situation where the contractor in front of the authority who do you think will win? [Laughs] So they see how we [they] can do sort of [be the] mediator. But our attitude is we shouldn't abuse such things.
>
> (Senior Technical Director, Design and Build contractor, Sports facility megaproject)

In both megaprojects, the clients demonstrated a keen awareness of the political and positional power inherent in their roles as authoritative entities – a government agency in the sports facility megaproject and the airport operator in the airport megaproject. This understanding enabled them to effectively facilitate regulatory approval processes or resolve issues smoothly. Given their status as either a government agency or airport operator, these clients garnered high levels of trust from various government and regulatory bodies, including the regulatory body governing building and construction activity, Civil Aviation, and fire, electrical, and plumbing authorities, a privilege not accessible to the broader project team. Consequently, the clients' recognition of their influential role and the power it entailed allowed the project team to engage in constructive deviance that ultimately benefited the projects.

Occasionally, there are instances where the project team must prompt clients for assistance, especially when encountering bottlenecks, as illustrated in the example above. The project team recognised the need to request aid through the client's legitimate position. However, it was crucial for the team to be conscious of not overlying on or abusing this support. This awareness reflects a balanced approach to leveraging the client's authority, ensuring that requests for assistance are made judiciously and responsibly.

Accelerating and leveraging collective knowledge and experience in decision-making

Given the fast-track nature of the decision-making environment of megaprojects, the ability to short-circuit processes through immediate access to a reservoir of knowledge and experience is critical. The ability to adeptly navigate the decision-making environment in megaprojects is significantly shaped by the collective knowledge and experience of the project team members. This manifests as a shared project memory and a pool of resources, which are continually enriched and expended by the team as the project evolves. This collective memory, comprising accumulated experiences and insights, serves as a vital asset in guiding decisions and strategies throughout the project's progression.

Box 9.6 Lesson – Continually acquiring and developing a stock of project knowledge and experience.

The varying skill levels and experiences of project team members matter significantly in how decisions are made. The importance of acquiring and developing a robust knowledge base and experience is evident in how it enhances their ability to navigate the complexities of the decision-making environment in megaprojects. This is demonstrated in the following quotes, which highlight the crucial impact of skills and experience on decision-making:

> It's a source of conflict. It was our fault. We didn't try to communicate it at the right time and at the right manner and we were late, and we might've caught people off guard…At that point maybe there were other things going on as well about costs. So, it got wrapped up into something else.
>
> (Senior Design Manager, Design & Build contractor, Sports facility megaproject)

> So, I think that's actually part and parcel of the entire process but that part where you feel like it's also about learning how to make your point across in a very reasonable and professional way so that they understand the value…
>
> (Associate Director, Architecture firm, Sports facility megaproject)

The quotes above reveal that project team members gradually developed a deeper understanding of tactics for collaboration. Instances of naivete or insufficient knowledge are cited as reasons for past setbacks, as illustrated by the reflection, "it was our fault. We didn't try to communicate it at the right time and right manner". This highlights the critical role that experience and informed judgment play in effective decision-making in megaprojects.

This growing understanding or knowledge about shaping situations to achieve favourable outcomes becomes a critical part of the project's organisational memory. It guides project team members on how to approach future project issues. For instance, the design manager honed specific communication skills tailored for working with the project team, mastering the art of knowing "the right time and right manner" to convey messages. Meanwhile, other team members gained more general skills, such as the ability to persuade more effectively, and learning to present points in a reasonable and professional manner. This accumulation of skills and experience not only enhances individual capabilities but also enriches the collective knowledge base of the project, aiding in more effective decision-making and problem-solving in future scenarios.

Crafting the unwritten code: establishing implicit "rules of the game"

For megaproject team members, there is typically an unspoken consensus on the expected behaviours of other team members, which they tacitly accept as the "rules of the game". These norms, while not always explicitly defined or easy to articulate, are underpinned by a specific kind of governmental rationality – governmentality. This concept is largely independent of formal instruction but still requires some level of intervention or participation.

The development of these "rules of the game" is a critical process that necessitates a deliberate investment of time and effort. It is essential to recognise that establishing these guidelines cannot be a passive or organic process. Instead, it requires intentional and structured organisation as an ongoing activity throughout the duration of the project. This proactive approach ensures that the rules evolve in a way that supports the project's objectives and aligns with the team's dynamics and challenges.

Box 9.7 Case study – The art of cultivating knowledge and collaboration in megaproject decision-making.

The development of necessary knowledge and understanding to work with each other on a megaproject does not occur instantaneously. As highlighted in the quotes below, this proficiency is cultivated through sustained interaction with the project team over time. The importance of frequent meetings, extensive exchanges, and generous sharing of information is emphasised. This process enables team members to establish effective methods of collaboration, which include exploratory approaches like "testing the water" and subtler strategies such as "sending things through the back door". These tactics are instrumental in shaping a cohesive working relationship and a mutual understanding among team members.

> You have to know the rules of the game…So it's like you play football you know. So, each one has to know the rules of the game because if you have to explain to the goalkeeper all the time then it's very difficult! So first you have to be sure that each one is informed of the rules of the game and the contract and after that to play without the need to go back to the rules and this requires a lot of meetings, a lot of exchange, and a lot of sharing. It'll be more at the beginning than with company where everyone will be used to work together.
>
> (Chief Executive Officer, Special Purpose Vehicle, Sports facility megaproject)

> our relationship, the pattern developed…that was 2.5 years ago. So now we've a way of working with each other. You test the water. You tend to try and send things through the back door first then you do the formal meeting. We try not to do the formal meetings until people are quite happy.
>
> (Senior Design Manager, Design & Build contractor, Sports facility megaproject)

> Sometimes you have to understand, for example, if you want to get something done that is not done it's how you react – do you bang tables or do you sarcastically do something or do you report to his boss? You have to be patient and explain to him why you need this urgently and if they don't do it then maybe you don't have to report first or should-should you report or should you find an alternative? Maybe you can talk to his subordinate to get an answer.
>
> (Chief Operations Officer, Special Purpose Vehicle, Sports facility megaproject).

These unspoken rules were regarded by project team members as unofficial principles that guided their decision-making. This understanding allowed them to navigate the project landscape more intuitively, enabling them to "play the game" without the constant need to explicitly refer to formal rules. This approach was notably mentioned by the Chief Executive Officer (CEO) of the Special Purpose Vehicle from the Sports facility megaproject, highlighting the reliance on these implicit guidelines for effective decision-making.

Similarly, the Chief Operations Officer outlined a range of tactics employed when seeking the cooperation of others, reflecting a strategic approach to problem-solving. These tactics included assertive measures like "banging tables" to signal urgency or seriousness, employing sarcasm to subtly convey a message, or escalating the issue to a superior like reporting to the individual's boss. Alternatives mentioned included seeking different solutions, like "finding an alternative" route to accomplish the objective, or engaging with subordinates, as in "talking to his subordinate", to bypass potential roadblocks. This variety in approach demonstrates a nuanced understanding of power dynamics and the flexibility required in navigating complex decision-making environments.

The role of personal relationships and networks

In their practice of either constructive or destructive deviance, project team members employ strategies and tactics to attain their desired goals, demonstrating their perception of the decision-making environment as akin to a game. Human resources in the form of personal relationships and networks were identified as the most frequently used by the participants from the two case studies to play the game.

Box 9.8 Case study – Cultivating personal relationships in professional contexts for smoother job execution.

Human resources, particularly in the form of specific project team members' social connections with influential individuals in positions of power, can facilitate the practice of deviance in project environments. These connections often provide access to insider knowledge and strategic alliances, which can be leveraged to navigate or circumvent standard procedures and rules.

> Even though I'm the COO I have to deal with the immediate level… You have to engage them informally as well and go out to lunch with them. When you go on holidays buy them some nice chocolates. When someone gives birth, you send them a congratulatory note and things like that. So, you try to interact with them at a personal level.
>
> (Chief Operations Officer, SPV, Sports facility megaproject)

> I deal actually before with all those people in other airports. We used to meet and socialise, not to discuss about work but we just "hi, hi" and – that is the first thing that makes the job easier.
>
> (Operational Readiness Airport Transfer Manager, Airport operator, Airport megaproject)

Success in professional settings, including the ability to gain support from others, often hinges on personal friendships and networks. As Brislin (1991) notes, it's a common phenomenon for deals to be made based on warm ties between individuals. This principle rests on the idea that people are more inclined to respond positively to requests from friends or those with whom they share close relationships. In light of this, participants felt the need to extend their engagement beyond professional boundaries, fostering personal connections through social interactions.

Success in navigating office politics requires cultivating relationships where others are invested in one's continued presence in an organisation (Brislin, 1991). Participating in social events and maintaining everyday cordiality are strategies that can be employed to nurture these relationships. Such engagements make it easier to garner care and support from others, ultimately facilitating smoother job execution. The existence of these social ties creates an assumption that enlisting support in various situations becomes more manageable, highlighting the value of personal connections in professional contexts.

The presence of pre-existing relationships is also a valuable asset in megaproject decision-making environments. These established connections can offer a foundation of trust and understanding, which is often crucial for efficient collaboration, smoother communication, and more effective negotiation within megaprojects.

Box 9.9 Lesson – Leverage established relationships to enhance trust and collaboration.

Friendships and social ties contribute significantly to megaproject environments through the trust and credibility that are built from having previously worked together. These established relationships often lead to a deeper understanding and mutual respect among colleagues, fostering an environment where individuals are more likely to rely on each other's expertise and judgments. This existing trust can expedite decision-making processes, enhance collaboration, and provide a reliable network for support and advice, which is especially valuable in complex or challenging project scenarios requiring rapid decision-making.

> Most of the authority are the same ones I worked with on the YYY. So, the person I dealt with in the LTA is the same one I dealt with here and people in MHA, Ministry of Home Affairs — they're the same people… I think it's very interpersonal with LTA. MHA its very much

> based on trust and having seen maybe how we operate and how we tend to deliver what we promise.
>
> (Senior Design Manager, Design & Build contractor, Sports facility megaproject)

> It makes the things far easier because you know you have this feeling of people trust you or they don't trust you or they know how it has happened before.
>
> (Technical Director, Design & Build contractor, Sports facility megaproject)

The quotes illustrate the significant role that prior relationships play in establishing credibility and trust. These pre-existing connections offer first-hand evidence of an individual's or a team's ability to fulfil commitments and deliver on promises. Colleagues who have witnessed past performances ("having seen how we operate and how we tend to deliver what we promise") gain confidence in their reliability and competence. This familiarity provides a strong foundation of trust, as stakeholders are reassured by their proven track record and are more likely to have faith in their capabilities and promises for future endeavours. This trust, rooted in past experiences and interactions, becomes a critical asset in facilitating smooth and effective collaborations.

Identifying and using cultural resources

Cultural resources in the context of megaprojects can be manifested in many ways.

Box 9.10 Case study – Using professional qualifications as a valuable cultural resource.

One significant aspect of cultural resources is the notion of legitimacy, which is often associated with individuals possessing specific credentials. For instance, being recognised as an architect bestows a certain level of legitimacy and respect, highlighting the importance of professional qualifications as a valuable cultural resource in these complex projects.

> If you are not somebody, you are nothing… they have to know who you are, "oh you are architect, OK, OK you are architect. I have to respect you. If you are not, I don't have to respect you". Like the contractor, they'll ask, "who are you?
>
> (Resident Architect, Client Design Consultant – architecture firm, Airport megaproject)

> So sometimes you have to put someone with architect status. They know this one I have to respect because his advice is final, and his say is final, and he has authority. If he says don't do something, it's don't do…when Tom goes people are looking at him and people start to note down what he's complaining but this project needs five Toms [Senior Architect]. So, it's too big and we need more stronger people to handle…sometimes when a more junior person goes on site they won't listen.
>
> (Senior Architect, Client Design Consultant – architecture firm, Airport megaproject)

The quotes highlight that the title of an architect carries a significant level of respect and authority, which can be strategically leveraged to influence and control the behaviour of others in project environments. In many Asian countries, the title of an architect is legally protected. Only individuals registered with each country's Board of Architects are entitled to use this designation. This regulation emphasises the importance and prestige associated with the role. In the airport megaproject, the presence of a senior architect on-site was crucial. Their status as a respected professional meant that their decisions were final and carried considerable authority. This respect for the architect's role and expertise ensured effective oversight and compliance on site, illustrating the significant impact of professional titles in shaping project dynamics.

Cultural resources can also encompass the collective competencies and expertise of individuals, rooted in each person's capability to provide a credible and reliable promise. This concept hinges on the recognition of collective skills and knowledge as integral components of a team's overall cultural capital.

Box 9.11 Case study – Leveraging credibility to gain support.

In the Sports facility case study, the credibility attached to different firms, practices, or methods was acknowledged as a valuable asset. This recognition played a crucial role in garnering support for their objectives, demonstrating how the reputation and reliability of entities and approaches can significantly influence project outcomes.

> …what we have at our disposal are our technical ways of showing something. So technical solutions are, although they're not real, they're theoretically proven so that's quite good in some ways because it's more black and white and to certain degree… to be able to show that something works or doesn't work.
>
> (Senior Design Manager, Design & Build contractor, Sports facility megaproject)

> But certainly the fire engineering solutions involves peer review process so the credibility of a well-established international practice outside of our team helps to support that the solution has been well thought through to get acceptance of it.
>
> (Senior Associate, Sports Architecture firm, Sports facility megaproject)

Equally significant is the ability to discern when to harness essential credibility. Given the complex nature of megaprojects, it becomes impractical to possess all the necessary competencies within one's immediate team. This is where the capacity to tap into or enlist additional resources and expertise becomes crucial in addressing the project's requirements. As exemplified in the second quote, the project team effectively utilised the credibility of a well-established international practice, beyond their core team, to garner support for their design solution.

It is also important to understand the advantages of strategically positioning oneself to enhance the effectiveness of persuasion when establishing credibility. This recognition highlights the significance of not only what is communicated but also how and by whom, in successfully influencing and convincing others in decision-making environments.

Box 9.12 Case study – Astute positioning to accelerate credibility establishment.

The former Project Director of the Sports facility megaproject reflected on a meeting he attended, where presenting himself in a specific manner was essential for gaining acceptance and trust from various stakeholders in prominent positions. This experience highlights the importance of perception and presentation in building credibility and establishing effective relationships in the megaproject environment.

> So, this was a very high-level meeting…and I'd only been working for about 2 months. They didn't know who this guy was, but the first thing was, "he's been in the country for 15 years, he's a PR". So that's the way that the CEO of the Sports government agency always introduced me. He was clever. He had to do that and then he'd say, "Robert brought the World Sevens to this country, Robert is Vice President of the national Rugby Union" and all that again and so people just went "ah, so he's one of us".
>
> (Former Project Development Director, Sports Facilities division, Sports facility megaproject)

The strategic positioning of the Project Director played a pivotal role in his rapid acceptance and trust-building among key stakeholders. By emphasising his status as a permanent resident who had lived in the country for 15 years and showcasing his impressive credentials, the CEO of the Sports government agency successfully presented him as "one of us", facilitating his immediate integration into the community. This integration was further solidified by his active role as the vice president of the national Rugby Union and his contributions to bringing the World Rugby Sevens Series to the country. These efforts were not just acts of good citizenship but also significant contributions to the country's sports development.

For the Project Director, especially as a recent appointee from a foreign country, it was crucial to be perceived as trustworthy and capable of managing stakeholder relations for a high-profile project like the Sports facility megaproject. This scenario is reflective of a common aspect of megaprojects, where teams comprising members from various international backgrounds collaborate on a temporary basis. Such projects often bring together diverse expertise and perspectives, necessitating effective strategies for trust-building and integration among team members from different cultures and countries. While trust-building typically takes time, there are ways for the process to be expedited, as this example of the CEO's strategic introduction of the Project Director has shown. This approach can be viewed as an effective shortcut for garnering support and achieving buy-in in a dynamic environment. In particular, astute positioning and presentation can significantly influence stakeholder perceptions and accelerate the process of establishing credibility.

Using deviance tactics to play the game

The megaproject decision-making environment is likened to a sort of tournament whereby victory is premised at least in part on the capacity of the project team members to practice deviance. Consequently, the outcomes of decision-making in these projects are largely a reflection of each individual team member's acumen and expertise. A variety of deviant tactics can be strategically employed to sway decision-making processes in megaprojects for both positive and negative purposes. It is crucial to discern when such actions are becoming disruptive and need to be curtailed, and conversely, when they are beneficial and should be encouraged. The ability to identify and act upon these distinctions is essential for maintaining the balance between facilitating progress and avoiding detrimental consequences in megaprojects.

The deviant tactic of *disrupting hierarchy* involves individuals intentionally disregarding established systems and processes, along with the associated lines of authority, respect, and duty (Ewick & Silbey, 2003). This approach is particularly disruptive because hierarchical deference typically operates on an unspoken acknowledgement of structural differences. By consciously ignoring these differences, individuals challenge the existing power structure, compelling it to assert itself more explicitly. Bypassing hierarchical levels can be a strategic means for individuals to achieve their objectives.

Box 9.13 Tactic – Disrupting hierarchy for innovation.

Project team members in the Sports facility megaproject disrupted the hierarchy in their approach to obtaining regulatory approval for the bowl cooling system, where they circumvented traditional hierarchical procedures to realise their goals.

> … to have that level of flexibility to think outside the box and to rethink again from first principles, one is I think it's not usually expected of them [those at the lower level of organization] so therefore I don't think that's necessarily that fair and at the same time on these kind of big scale off the wall kind of not standard projects you have to have some measure of thinking through first principles… inevitably, we have to talk to the directors of the agencies because the directors are the ones who will say OK they understand the benefit of this project and how it's being developed
>
> (Senior Project and Design Manager, Architecture firm, Sports facility megaproject)

Achieving flexibility to "think outside the box" and "rethink again from first principles" was only possible by approaching the top echelons of the organisation. Individuals in lower positions within organisations often find themselves constrained, limited to dealing with issues within the scope of their regular day-to-day operations. This limitation stems from the general expectation that such innovative thinking and fundamental re-evaluation are not typically in their purview. This highlights a hierarchical divide in organisational structure, where the freedom to innovate and challenge existing paradigms is often reserved for those at higher levels.

By straying from standard protocols and directly reporting issues to higher-ranking individuals, project team members managed to carve out a degree of agency and autonomy, which is often suppressed in a complex bureaucratic system. This approach allowed them to bypass the restrictive layers of bureaucracy, enabling a more direct and potentially effective problem-solving and decision-making. Such actions demonstrate how individuals within rigid systems can find ways to exert influence and navigate around procedural constraints to achieve their objectives.

Swapping deals involves the reciprocal exchange of favours, anchored in the mutual promise of access to shared benefits. The megaproject decision-making environment provides plenty of opportunities for compromises to take place and for the reciprocal exchange of favours. Achieving win-win situations, where different parties benefit in some way, is challenging. In the case of the Sports facility megaproject, a project with government association, the regulatory body faced a Public Relations challenge. The regulatory body had been working to prevent wasteful spillage of air conditioning in outdoor spaces. The bowl cooling system presented a dilemma: it was innovative but risked undermining the regulatory body's sustainability efforts. The solution: the regulatory body imposed a requirement that the system's energy be fully offset by renewable sources. This move served multiple purposes: it allowed the regulatory body to show the public that they were being equitable in their enforcement actions while still encouraging sustainable practices.

Box 9.14 Tactic – Swapping deals and reaching compromises.

The project team's effectiveness in using the "swapping deals" tactic is particularly notable in their efforts to reach a compromise for implementing the bowl cooling system. Mastering the art of compromise is essential in this context. It involves not only understanding and acknowledging the requirements of the other parties but also clearly demonstrating that these requirements have been considered in the decision-making process. A significant aspect of this involved the project team's skill in first "connecting all the dots" and identifying the "straight swap".

For the Senior Project and Design Manager, this represented a form of governance suited to the country's context. It demonstrated how authorities balanced innovation with public accountability. As he put it, it was a matter of "managing the concerns that they have…within the country's context of how governments run", maintaining public trust while ensuring fairness.

Project team members spoke of their skill in showcasing their interests favourably using different methods to secure approval or strategically influence others. *Rational persuasion* in the form of technical justifications was a common element in the toolkit that many project team members used. Team members from both case studies understood the power of strategic representation in megaproject decision-making. This technique involves presenting one's interests in a manner that is most likely to gain acceptance and influence others. Professional credibility in this context was seen not just as a testament to one's expertise but also as a tool that could be leveraged to advance hidden agendas. While general technical rationality may not always be sufficient to sway decisions, specialist knowledge can be a potent tool. For example, expertise in a niche area such as crowd control (see Box 9.15 below) can be used to justify certain decisions, such as design choices, that may otherwise be challenges. This tactic showcases how specialist knowledge can be strategically employed to align project decisions with specific interests or goals.

Box 9.15 Tactic – Rational persuasion: using credibility as a form of commodity.

Project team members often used information in a strategic way, sometimes to mislead or manipulate others for hidden goals. They frequently relied on rational persuasion, using technical logic or clear arguments to win support. This approach had two main purposes. First, as the Senior Design Manager on the Plaza canopy project explained, technical reasoning provided a clear, "black and white" method to show if a proposed solution would work or not. For them, this approach was a strong way to justify their arguments.

> ...what we have at our disposal are our technical ways of showing something. So technical solutions are, although they're not real, they're theoretically proven so that's quite good in some ways because it's more black and white and to certain degree... to be able to show that something works or doesn't work.
>
> (Senior Design Manager, Design & Build Contractor, Sports facility megaproject)

> So, a lot of explanations, a lot of sketches, modelling showing that at least the spaces the linking part is not compromised it's just not a direct route...learning how to make your point across in a very reasonable and professional way so that they understand the value of I think that's one of the thing that we'll always try and of course I won't say that its always successful
>
> (Senior Associate, Architecture firm, Sports facility megaproject).

The Senior Design Manager compared professional credibility to a valuable asset, similar to the trust people place in doctors. Their expertise in crowd control became a key tool to support their design choices. For instance, they argued for separate canopies instead of one large one at the Mass Rapid Transit station, claiming technical reasons, but actually, they just disliked the look of the large canopy. They convinced the client that multiple exits with side canopies were safer for fire and crowd control, making it hard for the client to disagree. Although this solution didn't fully protect against bad weather, it met other important goals like fire safety and managing crowds effectively.

It is, therefore, not uncommon for experts to use technical information to support their views. Being knowledgeable in specific areas gives them a sense of legitimacy and authority.

Gatekeeping refers to the strategic control over the flow of information. This practice includes intentionally withholding or filtering information as a means to exert power and influence over others. The ability of project team members to navigate the decision-making environment is largely influenced by their skill in communicating their message effectively. This skill is not just about the content of the message, but also about strategically managing how, when, and to whom the information is presented, all with the aim of garnering support for their objectives.

Box 9.16 Tactic – Using the gatekeeping tactic to champion a cause.

Project team members told many stories demonstrating their understanding of effectively using the gatekeeping tactic when championing a cause:

> We have to do a little bit of talking about what the expectations of what the authority is going to want to see…defining what messages we want to bring out now and what messages we can keep to later for a different audience…And you have to continually manage that to get people recognising that we've got this "package A" which we use for the authority and we've got this package B or C or D in some areas which is already advanced to a slightly different form.
>
> (Senior Associate, Sports Architecture Firm, Sports facility megaproject)

> Yeah, the way we do [it] is usually we do some consultation with first draft documents and we go to these consultations…After a few meetings, consultations [and] everybody agrees then we do a formal process where everybody signs off on it. There is the formal process, which is recorded through the document control, the software document but prior to that when the software document is being prepared, people are already aware of what they are going to receive.
>
> (Technical Director, Design and Build Contractor, Sports facility megaproject)

As seen from these quotes, the nature of the information, along with how often and when it is delivered, is carefully considered and well-planned. They highlight the careful orchestration of information dissemination by the design team to the regulatory body. The type of information provided, along with its frequency and timing, was meticulously planned to facilitate approval from the regulatory authorities. The stories show that it is important to prepare everyone involved about what to expect. The goal is to make sure people know what information they will receive, so they wouldn't be surprised. The main power in this situation is controlling what information is shared and when. The outcomes rely a lot on the skills and experience of the people involved.

Client's role

The client assumes a pivotal leadership role in megaproject decision-making, fostering cooperative behaviour among project team members. Occupying a central position within the power hierarchy, the client significantly influences the production and distribution of power on projects. This role is often a reflection of the client's political and economic decisions, making them a dynamic political entity in determining which interests are advanced on the project, thus displaying a decisive role in shaping project outcomes. The client, therefore, has the capacity to either facilitate or hinder deviant practices through the following ways.

Introducing and enforcing an appropriate ethical climate for decision-making

Clear rules and regulations are crucial for building trust and predictability in megaproject teams. Both clients and regulators wield significant influence in shaping the extent and nature of deviance within megaprojects through the ethical climate they establish. Furthermore, the mere existence of rules and regulations is insufficient; their consistent enforcement ensures that the intended governance structure functions as intended. If clients and regulators do not actively exercise their authority to intervene and prevent the normalisation of deviant practices, they inadvertently become part of the problem by allowing counterproductive behaviours to become accepted. Their active involvement is essential to ensure that projects maintain integrity, adhere to ethical guidelines, and achieve positive outcomes.

Proactively granting or withholding access to legitimacy and power, organisational resources, and the collective project memory

In numerous instances, clients, particularly those who are government agencies, have inherent access to valuable networks, as illustrated in the case study examples. These networks and resources can often be made available proactively, without awaiting a request. Taking the example of the bowl cooling system case study example, significant time and effort could have been saved if the clients had adopted a more proactive stance in granting access to their cultural capital and resources at an earlier stage. An anticipatory approach in sharing these resources could streamline processes and expedite project progress, avoiding unnecessary delays and inefficiencies.

Establishing a database of organisational resources and networks

Clients or their representatives have the ability to explicitly compile and maintain a knowledge inventory, akin to a "database" of resources for the project. This database serves to provide clarity on how individuals can access these resources and where to find them. This proactive approach can be initiated during the project's inception, at the procurement stage, and should continue throughout the project's lifecycle. By aligning resources with project requirements, it becomes possible to map out what is necessary and identify any gaps, which can then be addressed by referring to the database. This approach differs from the conventional practice of resource allocation during project progression. Instead, it advocates for the early availability of resources, ensuring that project team members are aware of their existence, know how to access them, and can obtain approval for their utilisation when deemed appropriate. This is essential given the typically fast-tracked nature of the megaproject decision-making environment.

Establishing and sharing the implicit "rules of the game"

Establishing unwritten guidelines for expected behaviours among team members is essential for smooth megaproject decision-making. It is important to emphasise that this process should be carried out explicitly and intentionally, rather than relying on the development of such guidelines organically. Proactive efforts in defining these expectations can help create a cohesive and productive team dynamic, ensuring that everyone is on the same page and aligned with the project's goals and values.

Facilitating or hindering deviant practices

A range of deviant tactics can be strategically used to influence decision-making, with potential outcomes that can be either positive or negative. The key lies in discerning when these actions become disruptive and need to be reined in, and conversely, when they are beneficial and should be encouraged to promote progress. The ability to make this distinction and take appropriate action is essential for maintaining a delicate balance that facilitates project advancement while preventing detrimental consequences in megaprojects.

Chapter summary

This chapter delved into the application of the governmentality framework in the context of megaprojects, highlighting the need to balance rules and enforcement with a degree of flexibility to encourage innovation and creativity. This chapter has offered a comprehensive exploration of the strategies involved in applying the governmentality framework in megaprojects, providing valuable insights for project professionals and researchers alike. The collection of examples and case studies allow readers to explore various governance and governing mechanisms suitable for their specific context to understand and leverage covert power and hidden influences.

References

Boyd, D., & Chinyio, E. (2006). *Understanding the construction client*. Oxford: Blackwell Publishing.

Brislin, R. (1991). *The art of getting things done: A practical guide to the use of power*. New York: Praeger Publishers.

Clegg, S.R. (1975). *Power, rule and domination.* London: Routledge and Kegan Paul.
Crawford, L., Cooke-Davies, T., Hobbs, B., Labuschagne, L., Remington, K., & Chen, P. (2008). Governance and support in the sponsoring of projects and programs. *Project Management Journal*, 39, 43–55.
de Certeau, M. (1984). *The practice of everyday life*. Berkeley: University of California Press.
Ewick, P., & Silbey, S. (1998). *The common place of law: Stories from everyday life*. Chicago, IL: The University of Chicago Press.
Ewick, P., & Silbey, S. (2003). Narrating social structure: Stories of resistance to legal authority. *American Journal of Sociology*, 108(6), 1328–1372.
Foucault, M. (1980). Body/power. In C. Gordon (Ed.), *Power/knowledge: Selected interviews and other writings, 1972–1977*. London: Penguin.
Garland, D. (1997). Governmentality and the problem of crime: Foucault, criminology and sociology. *Theoretical Criminology*, 1(2), 173–214.
Haugbolle, K., & Boyd, D. (2016). *Clients and users in construction: Research roadmap summary*. CIB Publication Vol. 2016, No. 408. Rotterdam: CIB.
Ninan, J., Mahalingam, A., & Clegg, S. (2019). External stakeholder management strategies and resources in megaprojects: An organizational power perspective. *Project Management Journal*, 50(6), 625–640.
Tzortzopoulos, P., Cooper, R., Chan, P., & Kagioglou, M. (2006). Clients' activities at the design front end. *Design Studies*, 27(6), 657–683.
Yukl, G., & Tracey, J.B. (1992). Consequences of influence tactics used with subordinates, peers, and the boss. *Journal of Applied Psychology*, 77(4), 525–535.

10 Summary and concluding thoughts

Summary of megaproject challenges

This book explores the impact of the nature of power relations in the megaproject decision-making environment. Chapter 1 identifies a crucial gap in the current understanding of megaprojects: the complex nature of power relations that influence decision-making. The chapter critiques existing governance frameworks and models for their emphasis on prescribing ideal behaviour, arguing that they fall short in addressing the diverse and complex forms of power at play in megaproject decision-making environments.

In Chapter 2, the exploration of megaproject discourse revealed that the research community has predominantly focused on three principal explanations for poor performance in megaprojects: strategic rent-seeking, governance regimes, and organisational complexity. Despite their differences, these approaches share a common underlying assumption: the concept of actor "farsightedness". This notion suggests that project governance can be proactively and effectively designed in anticipation of future events and challenges.

This perspective often leads to the belief that issues in megaprojects primarily stem from inadequate planning and administration. It implies that if only plans and policies were more comprehensively conceived and executed, based on rational and democratic arguments, many problems encountered in megaprojects could be avoided or mitigated. This view paints a scenario where positive change is achievable through the meticulous application of such well-considered plans and policies. Arguably, the traditional functionalist and instrumental project management approaches and models have not provided adequate insight into the social and situational construction of megaproject decision-making practice.

The narrative of decision-making in megaprojects is far from straightforward; it is a complex tapestry woven with compromise, accommodation, uncertainty, contradiction, and a web of partially implemented policies marked by deviations and nonconformities. This complexity is a fundamental characteristic of megaprojects, challenging the notion of strict adherence to predefined protocols and processes.

Previous research into the management of deviations in megaprojects revealed a pragmatic approach adopted by spontaneously assembled teams. Their focus often leans more towards "getting the job done" rather than rigidly following a set of

DOI: 10.1201/9781003377481-10

predetermined guidelines. While formalised, legitimate forms of power structure the day-to-day practices, they coexist with more covert forms of power. These less visible elements include political manoeuvres and strategic calculations that are often, equally, or in some cases, more influential in shaping decision-making processes. These hidden layers of power dynamics play a critical role in the actual unfolding of megaprojects, influencing outcomes in ways that formal structures alone cannot account for. Therefore, to truly understand and effectively manage megaprojects, it is essential to gain an explicit understanding that extends beyond the formally prescribed mechanisms to encompass the informal, often unseen, power relations that significantly influence decision-making.

The review within the megaproject discourse illuminated crucial implications for this book. A notable gap identified was the need for an in-depth exploration of the spontaneous governing practice of clients and project team members, focusing on the actualities of megaprojects as they progress. This book aims to shift the focus from conventional approaches that typically concentrate on the prelaunch planning or postcompletion analysis of projects. Instead, it delved into and unravelled the diverse forms of power that are exercised in the midst of megaproject management.

A governmentality framework for megaproject decision-making

Chapter 3 introduces an alternative way of thinking about power, centred around the concept of governmentality, as developed by the social theorist Michel Foucault in the 1970s. This concept was proposed as a more effective means of exploring power relations within the context of megaproject decision-making. Foucault's concept of governmentality, emerging from his profound investigations into political power, provides a useful language and analytical lens for examining the dynamics of power in megaproject governance, encompassing both the macro spaces of governance and the more confined, everyday workplaces of clients and project team members.

In this framework, power is conceptualised as a form of guidance or an attempt to shape behaviour with a certain level of intent and deliberation, aimed at achieving a variety of ends. This understanding of power extends beyond coercive force or authoritative command; it encompasses subtler forms of influence and direction. One of the key strengths of applying the governmentality approach to megaprojects lies in its capacity to delve into the "messy realm" of real-world practices and relations. It provides a lens to understand the compromised, corrupted, and realities of how governance mechanisms are partially enacted in the complex and often unpredictable environment of megaprojects.

A governmentality framework for decision-making is outlined in Chapter 4. The framework offers a distinctive and insightful approach to studying the governing processes of megaprojects, focusing on power relations rather than on power as an entity in itself. This approach aligns with Foucault's conceptualisation of power as a pervasive, circulating force, rather than a static commodity possessed by certain entities or individuals. Power, in this view, is ubiquitous and dynamic, functioning like a chain with constant movement and exchange across various nodes and interactions.

The orientation of governmentality work, particularly its emphasis on exemplary criticism, facilitates a more open-ended and exploratory analysis of power. It steers clear of prescribing fixed principles or ideologies for governance. Instead, it encourages posing critical questions about power, its mechanisms, and its effects, without seeking to establish a normative framework for governing. This approach is particularly suited to the decision-making environment of megaprojects, allowing for the mapping of the diverse and often complex ways in which governance mechanisms are enacted in the real world. Such an approach moved beyond idealised typifications to capture the nuanced realities of megaproject governance.

The governmentality framework's perspective on power analysis considers that the exercise of power through a range of deviant tactics is not inherently negative. Contrary to traditional views that often perceive power as predominantly coercive or manipulative, the framework acknowledges that the practice of deviance can lead to both positive and negative outcomes. It recognises the potential for deviant tactics to contribute constructively to project outcomes, as well as the risks they pose.

Key learnings

Chapter 4 presents an in-depth analysis of two case studies of megaprojects, providing a rich exploration of the real-world decision-making environment within such ventures. The chapter vividly illustrates various instances of deviations from established norms and protocols through multiple examples drawn from these case studies. These examples highlight the stories of power as narrated by individual clients and project team members, shedding light on the often hidden and mundane everyday practices that are used to challenge, resist, or reinforce the formalised structures and protocols of megaproject governance.

Stories told by clients and project team members were dissected, focusing on their everyday acts of deviance in the context of megaprojects, allowing for a deeper understanding of the taken-for-granted aspects of how they navigate their roles and responsibilities. These narratives revealed the complexities of real-world project management, where formalised governance frameworks, though clear in defining roles and prescribing processes, often fall short in addressing the practical challenges encountered on the ground.

In Chapter 4, deviant tactics are categorised into two distinct types: constructive and destructive deviance. The chapter also includes stories shared by megaproject clients and project team members, highlighting their awareness of the complex and fluid nature of power relations within megaprojects. They acknowledged that while institutionalised systems and protocols set certain expectations and limitations, they also provide openings for the strategic mobilisation of resources to achieve more advantageous results. This recognition emphasises the duality of such governance structures: they are both constraining and enabling.

In Chapters 5 through 8, the book delves into the intricate world of megaproject decision-making, framed within a governmentality perspective. This approach reveals an environment where rules, rights, roles, and responsibilities are not just binding constraints but also elements of a complex game with its own internal logic and legitimacy. Like the rules of a game, these elements provide a structure within which the megaproject operates, offering both limitations and opportunities for those involved.

Megaprojects are characterised by a dense network of formal rules and expectations, typically embodied in contracts and official documentation, which prescribe specific behaviours and practices. However, the experience of working on these projects often leads to the perception that many of these rules, protocols, and procedures are not immutable laws but rather constructs that are historically and socially shaped. As a result, these elements are seen as open to interpretation, susceptible to change, and potentially manipulable for self-interest. Chapters 5 to 8 present various facets of this governmentality framework, illustrating how it applies to the decision-making environment in megaprojects. These chapters explore the nuanced ways in which clients and team members in megaprojects interact with the established governance structures, highlighting the flexibility, adaptability, and agency that individuals and groups exert within the confines of the formal rules. The exploration of megaproject deviance has yielded several key learnings that significantly enhance our understanding of decision-making in these large-scale projects.

Governmentality is a realistic approach to decision-making in megaprojects

The governmentality framework applied to megaproject decision-making highlights the necessity of a balanced integration of two critical factors: governance and governing. This framework reveals that megaproject clients and team members often navigate a paradoxical landscape. On the one hand, they are ideologically aligned with and operate within the formal governance frameworks, roles, and relationships established by their professional training and values. On the other hand, they are situated within a practical topology of power that is oriented towards the pragmatic reality of "getting the job done".

This dichotomy emphasises the need to acknowledge and accept the coexistence of formalised governance mechanisms and informal governing practices within megaprojects. Both these orientations are essential, and the design of governance frameworks for megaprojects must accommodate these dual rationalities. The formal structures provide necessary guidelines and expectations, while the informal practices offer flexibility and adaptability to address the often unpredictable challenges of megaproject execution.

Embracing the governmentality approach to megaproject decision-making has profound implications for the management of these projects. Recognising the existence and practical importance of informal governing practices is crucial. It allows for a deeper understanding of the actual decision-making environment, including the spaces where deviations and adaptations occur. This recognition is the first critical step in efforts to guide and influence future megaproject decision-making towards improved performance. By accepting and incorporating the realities of both formal governance and informal practices, clients, project team members, and policymakers can develop more effective and responsive strategies for more successful outcomes.

The practice of deviance in megaprojects is undertaken in a continuum

This book has revealed that the reality of decision-making in megaprojects is far from the structured and orderly ideals often depicted in governance frameworks. One of the key insights is the inherent generality of governance policies and frameworks in megaprojects. These frameworks are typically composed of broad principles designed to be universally applicable across various aspects of the project, regardless of their specificity or uniqueness. This overarching generality results in a rule-based system that is inherently open to interpretation and adaptation.

As a result of this generality and subsequent interpretative nature of rules, deviations from prescribed protocols and procedures are not anomalies but rather a regular occurrence in the management of megaprojects. Clients and project team members often resort to these deviations as a means to resolve issues, navigate challenges, and progress the project effectively.

While the situations that give rise to these challenges can vary greatly, the responses and resolutions employed by clients and project team members are not without patterns. This book has revealed that clients and project team members typically draw from a specific repertoire of tactics and forms of power. These responses can be broadly categorised into two types of deviance: constructive and destructive.

A key learning achieved is the placement of megaproject deviance on a continuum. This continuum acknowledges that there are acceptable and, at times, beneficial levels of deviance. Constructive deviance, in particular, can be advantageous for megaprojects, offering innovative solutions and adaptability in the face of complex challenges. This understanding challenges the traditional notion of project success and failure, which often hinges on strict adherence to predetermined objectives and standardised protocols.

This book suggests that the effectiveness of a megaproject should not be solely judged based on its conformity to initial plans and norms. Instead, a more realistic and comprehensive approach to evaluating project performance should consider the different levels and types of deviant behaviour. By recognising and understanding these behaviours, clients, project managers, and other stakeholders can better facilitate constructive deviance, which can lead to innovative and effective solutions. At the same time, this approach allowed for the identification and interruption of destructive behaviours, thereby enhancing overall project performance.

Factors facilitating deviance

The practice of deviance in megaprojects, often manifested through subtle and concealed practices, is influenced by a range of enabling factors. These factors not only determine the extent and nature of the deviance but also the degree to which it is constructive or destructive. Understanding these influencing factors is vital in the context of megaproject decision-making. Such an understanding is crucial for promoting constructive deviance and mitigating destructive behaviour, thereby steering projects towards enhanced performance. Acknowledging and addressing these underlying factors can lead to more effective strategies that capitalise on the potential benefits of deviant practices while minimising their adverse impacts.

1 *Ethical climate*: The role of the client and other key stakeholders in construction projects is crucial in shaping the ethical landscape of a project. When clients or influential figures overlook questionable practices, these actions can gradually become accepted and eventually become the norm. This phenomenon, known as the normalisation of deviance, isn't just influenced by what happens inside an organisation. It is also shaped by the external environment, such as industry regulations and societal expectations (Hajikazemi et al., 2020). For instance, despite having standards and regulations like the UK Corporate Governance Code and the Government Prompt Payment Code, if regulations do not enforce these rules strictly, undesirable practices can become commonplace. This highlights the importance of considering both internal factors (like clients and stakeholders) and external ones (such as regulatory bodies and community groups) in understanding and addressing deviant behaviours in projects.
2 *Competitive project business environment*: In the fiercely competitive construction project environment, it is common for companies to submit unrealistically low bids to secure projects. This practice, often driven by economic pressures, can have serious consequences. As Flyvbjerg (2005) points out, this "strategic representation", or the use of misleading information for political gain, leads to projects starting with slim, often decreasing profit margins. This financial strain pushes stakeholders to seek ways to improve their financial standing, often at the expense of project goals and ethical standards. The priority placed on financial success over project objectives and ethical considerations raises critical questions. This trend has been extensively studied particularly how different procurement methods influence this competitive environment (e.g. Jefferies et al., 2013; Teo et al., 2012).
3 *Autonomy*: The concept of autonomy in construction is a double-edged sword. On one hand, autonomy is essential for fostering innovative behaviours, such as creative problem-solving. On the other hand, it can also open doors for unethical actions. This darker side of autonomy is often termed Machiavellianism, described by Moore et al. (2012) as the tendency to manipulate and be ruthless for personal gain. Stakeholders with a high degree of autonomy in decision-making may use their freedom to engage in destructive deviance, manipulating situations for their own benefit (Yildiz et al., 2015).

Recommendations

Three key recommendations to improve the megaproject decision-making environment are now provided.

Identifying appropriate decision-making contexts relevant to project stages

The governmentality framework for megaproject decision-making offers a comprehensive lens for understanding the context in which megaproject decisions are made. A key aspect of this framework is the identification of the entrepreneurial quadrant as a particularly conducive context for decision-making in megaprojects. This quadrant represents a blend of formal and informal governance approaches, fostering an environment where innovative and adaptive decision-making can thrive.

However, it is crucial to note that no single decision-making environment, including the entrepreneurial quadrant, can universally address all the challenges inherent in megaprojects. Each megaproject is unique, with its specific complexities and contextual factors, meaning that different challenges may require different decision-making approaches. As such, the entrepreneurial quadrant is not a one-size-fits-all solution, but rather one conducive context among others.

The framework further reveals that the interplay between the formal and informal aspects of governance and governing leads to four behavioural contexts for decision-making:

1 *Entrepreneurial*: characterised by innovation, flexibility, and adaptability, this context encourages creative problem-solving and strategic thinking, often blending formal rules with informal practices.
2 *Box ticking*: focused on compliance and adherence to formal rules and procedures, this context may prioritise process over outcome, often leading to rigid and bureaucratic decision-making.
3 *Deception*: involving manipulation or strategic misrepresentation, this context can lead to decision-making that prioritises specific interests or agendas, sometimes at the expense of the project's broader objectives.
4 *Recklessness*: characterised by a disregard for formal rules and a lack of enforced structured governance, this context can result in hasty, ill-considered decisions that may jeopardise the project.

Each of these quadrants possesses distinct characteristics that influence how team members may engage in deviant practices. The understanding of these behavioural contexts is crucial for predicting decision-making behaviour within specific environments of megaprojects.

Moreover, the application of specific behavioural contexts to different stages of a megaproject can be a strategic approach to enhance overall project performance. This strategy recognises that different phases of a project may benefit from different decision-making styles and governance approaches.

For instance, the entrepreneurial quadrant could be particularly valuable during the initial stages of a megaproject, such as planning and design. During these phases, flexibility and out-of-the-box thinking are crucial for developing effective strategies, overcoming unforeseen challenges, and setting the foundations for the project. Conversely, the box-ticking quadrant might be more suitable during the project delivery phases. In these stages, ensuring compliance with regulatory standards, safety protocols, and quality benchmarks is critical. Therefore, it is recommended that before the commencement of a megaproject, clients and project team members should carefully consider and identify the behavioural contexts most appropriate for each stage of the project.

Establishing mechanisms to support conducive megaproject decision-making

The rigorous enforcement of formal governance mechanisms, as discussed in Chapter 6, can lead to unintended consequences. When disciplinary power is applied stringently, it may drive project team members towards deceptive practices. On the other side of the spectrum, a decision-making environment with loosely enforced rules, as explored in Chapter 8, can be conducive to innovative and entrepreneurial behaviour. However, this approach also carries inherent risks. Chapter 7 delves into the danger of questionable practices becoming normalised within a project environment, especially when they are tolerated or overlooked by those in positions of power. Early warning systems need to be implemented to detect and address potentially harmful self-interested behaviours before they escalate or become ingrained in the project culture.

Understanding that each quadrant in the decision-making framework has its own set of advantages and disadvantages is crucial for effective megaproject management. This recognition allows for the identification of specific interventions needed to promote desired decision-making behaviours at various stages of the project. Here are some pertinent questions that arise in relation to this understanding:

How can we tailor interventions to align with the unique characteristics of each quadrant?

When is it appropriate to use disciplinary mechanisms and conversely when is it appropriate to use nondisciplinary mechanisms in megaproject decision-making?

What are the most effective mechanisms to transition between quadrants as a project evolves?

What specific training or support can be provided to project team members to navigate different decision-making environments effectively?

How can early warning systems be designed to detect when the decision-making environment is veering towards the destructive ends of the spectrum?

Within this context, the client holds a pivotal role as a change agent, particularly in establishing and nurturing the appropriate governance context that fosters and encourages desired decision-making behaviour. The influence of the client in megaproject development and delivery cannot be overstated; they are a critical component whose actions and internal dynamics significantly impact the project's success. Several aspects of the client's role in this context are the following:

1 Introducing and enforcing an appropriate ethical climate for decision-making
2 Proactively granting or withholding access to legitimacy and power, organisational resources, and the collective project memory
3 Establishing a database of organisational resources and networks

4 Establishing and sharing the implicit "rules of the game"
5 Facilitating or hindering deviant practices.

Developing thresholds for constructive and destructive deviance

The recognition that deviance in megaprojects exists on a continuum, ranging from highly destructive to highly constructive, emphasises that not all deviations from prescribed norms are detrimental. In fact, constructive deviance can be a valuable asset to megaprojects, fostering innovation and creativity that can significantly enhance project performance. However, the key is to manage this deviance judiciously, ensuring that it contributes positively to the project objectives while maintaining overall project integrity and compliance with essential standards.

The categorisation of deviant practices into constructive and destructive types is particularly useful for understanding the complex dynamics within megaproject environments. The recommendation to examine the levels and types of deviance specific to a project before its commencement is a strategic approach that can inform the design of more effective governance frameworks. Some key considerations for implementing this approach include the following:

1 Prior to the start of a project, a thorough analysis should be conducted to understand potential deviant behaviours that might occur. This involves assessing the project's context, the stakeholders involved, and historical patterns of behaviour in similar projects.
2 To effectively harness deviant behaviours, it is crucial to categorise types of deviant tactics. First, the categorisation of constructive deviant tactics into "critical" and "noncritical" groups allowed for the prioritisation of behaviours that are essential and most beneficial to the project. Similarly, it is equally important to discern and categorise types of destructive deviance to distinguish between unacceptable and, in rare cases, acceptable behaviours.
3 Define appropriate thresholds to distinguish between acceptable and unacceptable deviant behaviours. These thresholds should be clear, well-communicated, and understood by all project participants.
4 Identify contexts where constructive deviance could be beneficial and establish mechanisms to encourage and support such behaviour.
5 Develop strategies to monitor and mitigate destructive deviant behaviours. This includes establishing clear consequences for actions that fall outside acceptable boundaries and creating clear robust oversight mechanisms.
6 Implement training programs for project team members to understand the types of deviant behaviours, their potential impacts, and the governance frameworks in place. This will ensure that all team members are aware of the expectations and boundaries related to deviant behaviours.

Chapter summary

This chapter concludes the book by emphasising the significant role of power relations in megaproject decision-making. Traditional governance models were critiqued for their inability to capture the complex, often informal, dynamics of power in these projects. A focus on governmentality theory, as proposed by Michel Foucault, provided a new lens to analyse these power dynamics, highlighting the interplay between formal governance and informal governing practices. This book delves into the "black box" of megaprojects, uncovering the often-unseen aspects of decision-making practice through the examination of everyday, informal negotiations and interactions. The stories presented in this book revealed the critical role of multi-level networks in problem-solving and decision-making processes within the megaprojects. The narratives showcased how power is exercised in seemingly mundane and routine activities, pivotal to decision-making. These stories highlighted a variety of deviant tactics used by clients and project team members to navigate the complex decision-making environment. These tactics were employed to reinforce, influence, manipulate, challenge, or resist the formalised structures and protocols of megaprojects, thereby providing new insights into the hidden dynamics of megaproject practice. The chapter concludes with practical recommendations for identifying and managing different decision-making contexts, tailoring governance frameworks, and fostering an environment conducive to effective and ethical decision-making in megaprojects.

References

Flyvbjerg, B. (2005). Design by deception: The politics of megaproject approval. *Harvard Design Magazine*, 22, 50–59.

Hajikazemi, S, Aaltonen, K., Ahola, T., Aarseth, W., & Andersen, B. (2020). Normalising deviance in construction project organizations: A case study on the collapse of Carillion. *Construction Management and Economics*, 38(12), 1122–1138.

Jefferies, M., McGeorge, D., London, K., & Rowlinson, K. (2013). Relationship contracting: A case study of the Top Ryde Public Private Partnership (PPP). In *Proceedings of the 19th CIB World Building Congress*, Brisbane, 5–9 May.

Moore, C., Detert, J., Trevino, L., Baker, V., & Mayer, D. (2012). Why employees do bad things: Moral disengagement and unethical organisational behaviour. *Personnel Psychology*, 65, 1–48.

Teo, P., Bridge, A., Gray, J., & Rowlinson, S. (2012). Towards testing a new first-order decision making model for the procurement of public sector major infrastructure. In *Joint CIB W070, W092 & TG72 International Conference: Delivering Value to the community*.

Yildiz, B., Alpkan, L., Sezen, B., & Yildiz, H. (2015). A proposed conceptual model of destructive deviance: The mediator role of moral disengagement. In *11th International Strategic Management Conference 2015*, Procedia – Social and Behavioural Sciences 207, 414–423.V

Index

Note: **Bold** page numbers refer to tables and *italic* page numbers refer to figures.

For Product Safety Concerns and Information please contact our EU
representative GPSR@taylorandfrancis.com
Taylor & Francis Verlag GmbH, Kaufingerstraße 24, 80331 München, Germany

www.ingramcontent.com/pod-product-compliance
Ingram Content Group UK Ltd.
Pitfield, Milton Keynes, MK11 3LW, UK
UKHW020005200926
475702UK00015B/2488
* 9 7 8 1 0 3 2 4 5 5 4 5 7 *